Rubbish!

Human animals, so I'm told
Are superior beings, just look, behold,
No lowly animal left this mess
It's made by man, whom God did bless
With working hands, and an active brain
His epitaph here, a despoiled terrain
Man, look hard, —then bow in shame
Clear up your rubbish and honour your name.

—By Gweneth Mary White

Take responsibility for making a change. Don't wait for somebody else. Don't base your actions on what other people do, but on what you know to be right and good. If changes, then be a part of that change.

垃　圾

人类动物，我得人相告，
属高等生灵，可你瞧瞧，
这烂摊非低等者造次，
赖上帝给予人类所赐
劳作之手及活跃之脑：
墓志铭在此，宛如遭盗。
人啊，细瞅再惭愧低头，
捡拾起垃圾，不辱名由。

——温尼斯·玛丽·怀特

担负起改变之责。切勿等待他人。请把你自己的行动建立在你知道的是非的基础之上，而不是他人行动的基础之上。如果这发生了变化，那就成为该变化的一部分吧。

[澳]伊凡·戴维斯 著　董淑铭 译

对话：
为了更美好的明天

DIALOGUE
FOR
A BETTER
TOMORROW

浙江工商大學出版社
ZHEJIANG GONGSHANG UNIVERSITY PRESS

图书在版编目(CIP)数据

对话：为了更美好的明天：英、汉 /（澳）戴维斯(Davis,I.)著；董淑铭译. —杭州：浙江工商大学出版社，2015.4(2015.9 重印)

ISBN 978-7-5178-0967-8

Ⅰ. ①对… Ⅱ. ①戴… ②董… Ⅲ. ①环境保护—普及读物—英、汉 Ⅳ. ①X-49

中国版本图书馆 CIP 数据核字(2015)第 061460 号

对话：为了更美好的明天

[澳]伊凡·戴维斯 著　董淑铭 译

责任编辑	沈　娴
封面设计	王好驰
责任印制	包建辉
出版发行	浙江工商大学出版社 (杭州市教工路 198 号　邮政编码 310012) (E-mail：zjgsupress@163.com) (网址：http://www.zjgsupress.com) 电话：0571-88904980,88831806(传真)
排　　版	杭州朝曦图文设计有限公司
印　　刷	浙江云广印业有限公司
开　　本	880mm×1230mm　1/32
印　　张	6.875
字　　数	164 千
版 印 次	2015 年 4 月第 1 版　2015 年 9 月第 2 次印刷
书　　号	ISBN 978-7-5178-0967-8
定　　价	30.00 元

浙江工商大学出版社营销部邮购电话 0571-88804970

'The idea to start picking up rubbish' came from the fact that I knew I had to do something about the anger I was feeling. Dealing with my anger intellectually was not enough. I had to take action.

"捡拾垃圾的想法"来自我觉得必须做些什么来平息自己的愤怒情绪。理性地控制我的愤怒还不够。我得采取行动。

I don't always look happy when I am picking up rubbish.

捡垃圾时，我的表情并不总是很开心的。

This is on Hongyanjian Peak of the Da Luo Mountain area. Along with a couple of friends, we picked up most of the rubbish in this little area in less than an hour. On another couple of days we finished the job. Follow up maintenance will be required for as long as the mountain is there. I believe there will always be a few people amongst us who will drop their litter if nobody else is watching, even in a beautiful place like this.

此照片拍摄于大罗山上的洪岩尖。在几个朋友的帮助之下，不到一小时我们便清理了这个山头上的大多数垃圾。随后几天我们彻底清除了这个地方的垃圾。只要这个山头在，就需要后续的维护工作。我相信如果没有人监督，总会有人在像这样美丽的地方扔垃圾。

At times we have reverted to using plastic bags to collect rubbish because the rubbish collection points often have bins that are overflowing with garbage which then spreads out making a mess. At least we can leave the rubbish contained and tidy in disposable bags.

有时候我们还是会用塑料袋来装垃圾，因为垃圾收集点的垃圾箱常常都装得太满，以至于垃圾散落四周，一团糟。至少我们能将垃圾装袋，便于搬运。

An example of litter accumulating in an otherwise pretty place.

垃圾在一个原本应该美丽的地方累积。

It starts with maybe one piece of paper tissue carelessly discarded and then it quickly becomes yet another place to dump rubbish. Rubbish attracts rubbish. It keeps growing to a hill.

这也许始于一张不小心丢弃的纸巾，之后这便成了另一个垃圾倾倒场。垃圾招来垃圾，不停地堆积，成为一座山。

This rest area and shelter hut does not have rubbish bins or paid workers to clean it. Is that really any excuse for this? We can never get enough workers to pick up our rubbish if we drop it everywhere. No government in the world has enough money to do that because people have gone into every corner of our little world and we continue to do so. I have been to some really obscure and secret little places even here in Zhejiang Province. People drop rubbish even in those places and they will continue to do so until our changed behaviour becomes a long-term habit.

这个休息亭没有垃圾箱，也没有工人去清理。这是真正的理由吗？如果我们随处乱扔垃圾，我们永远不会有足够的工人来收拾这些丢弃的垃圾。世界上没有任何一个政府有足够多的钱来处理这件事，因为人们已涉足我们小小地球的每一个角落。即便在浙江我曾到过的一些真正不为人知的小地方，人们也在那些地方扔垃圾。在不乱扔垃圾的行为成为常态之前，他们会继续乱扔。

The shelter after clean-up.

收拾干净后的休息亭。

A few workers among many people. This photo was taken in the early days. The bank of the reservoir does not have rubbish on it now such as you can see in this photo. Consequently the water is also mostly free of floating rubbish.

游人中有几个工作者。此照片摄于垃圾捡拾工作的早期。如今，正如你所见，水库岸边已见不到垃圾了。因此水面上也不大看得见漂浮的垃圾。

Euken and Catherine. Catherine was the first person to stick with me and help consistently for a long time. Euken came purely because he was passionate about the environment.

于肯和凯瑟琳。凯瑟琳是第一个跟随我工作，在很长的一段时间里不断给我帮助的人。于肯因为对环境的热爱而加入我的行列。

Betty (张玲燕). Betty was by far, the longest serving and most consistent of any of my regular helpers and there have been many of them.

贝蒂（张玲燕）。迄今为止，贝蒂是我所有助手中“服役”时间最长、最持之以恒的人。还有其他很多类似的人。

This is Sam (温正义) getting to work in a really nasty situation. There is nothing pleasant about cleaning up these piles of garbage that accumulate. How do we do it? We laugh and make jokes, and then when we finish, we take a moment to see how different the environment looks and we can smile with a sense of accomplishment.

萨姆（温正义）在收拾一堆恶心的垃圾。要收拾这些逐渐累积的垃圾堆，不是一件愉悦的事。我们怎么对付过来的？我们边开玩笑边工作。当我们收拾完之后，停下来看看我们的劳动成果，就会满足地微笑。

Shannon (俞铁樱) may not like her photo being taken but she never let a rainy day stop her coming to work.

香农（俞铁樱）不喜欢照相，但她从来没有因为下雨而缺席捡拾垃圾的活动。

This photo reminds me of some of the incredible characters that I have met. There is a young lady in the foreground of this photo who was very shy about having her photo taken. Her name was Sky and it was always a pleasure to work and joke with her.

这张照片让我想起我所遇到的那些令人难忘的人，照片上最前面的那位女士很羞于照相。她叫“斯凯”，与她边工作边开玩笑总是很愉快的。

There are exceptions to every general statement but I still find myself saying sometimes that girls are the best workers.

凡事都有例外，但有时我还是要说女孩子是最好的工作者。

This is a pretty normal group from Wenzhou University, taken as we walk to work.

这是温州大学常见的一个团队，摄于我们去劳动的路上。

There have been a lot of group photos taken. This one does not include me and I put this one here specifically because we need to remember sometimes that there are many other groups and individuals working on environmental protection projects in many different ways.

有很多小组集体照。我不在这一张里，我把这一张照片放在这里，是因为有时我们需要记住还有很多团体和个人在以不同的方式为环保事业奋斗着。

A different group. Sometimes different groups would join me for one time or a few times and then go off to other activities. Most people in this group were from Wenzhou Medical University.

一个不同的团队。有时候不同的团队会跟我一次或者几次，之后就去参加其他活动。这个团队的大多数人来自于温州医科大学。

Sometimes when one is doing work like this, the task can seem too big. I set a limit to which we would work and that was, if we could see rubbish from the track then we would pick it up if we could. There is a simple motto here to live and work by: Do what you can do.

有时候需要收拾的范围非常大，我规定了我们工作的范围，那就是如果从小道上能看见垃圾，那就尽力将其捡起来。在这里生活和工作的座右铭很简单：做你能做的。

Rope work. In this situation we chose to pull large bags of rubbish up onto the walkway using a rope.

绳索工作，即在此情形下，我们用绳索将一大袋一大袋垃圾吊到路面上来。

I don't know a lot about working while hanging off a rope but we have cleaned up a few places using this technique.

我不记得有多少次挂在绳子的另一端工作，但我们确实使用这种方法清理了一些地方的垃圾。

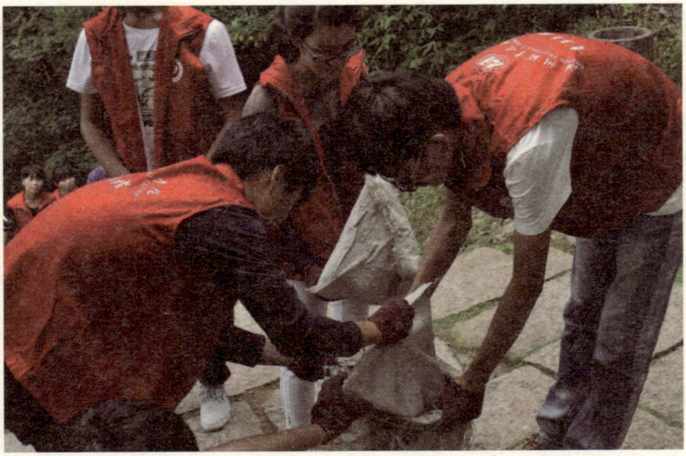

Usually sorting rubbish for us was an exercise in getting the most work output we could from the people we had on that day. Sometimes people would suggest that we should gather and sell the bottles we collected but I have always resisted that idea for two reasons. The first is that we have work to do and collecting bottles to make money would distract us from our main task. The second reason is that there are some people in our society that depend on an income they can get from selling bottles. They need that income more than me so I am not going to take it from them.

通常，垃圾分类是将当天捡到的垃圾进行合理分工的一个过程。有时人们建议我们把收集到的瓶子卖掉，但我拒绝了这个主意。原因有二：其一，我们有工作要做，收瓶子去卖钱有悖我们的主要目的；其二，社会上有些人依靠收集瓶子卖钱为生，他们比我更需要那笔钱，因此我不会从他们手里抢夺这份收入。

Getting ready for the descent. It is not easy work carrying bags of rubbish down from the mountain so it always takes a while to get organised.

准备下山。要将垃圾带下山并不容易，所以总是得花一些时间来组织分工。

The shoulder pole is an excellent way to carry heavy loads of trash off the mountain and into the places where it can be processed and disposed of correctly.

扁担是一种很好用的工具，用于将很重的垃圾挑下山，送到处理场进行妥善处置。

Carrying bags of rubbish off the mountain.

将垃圾带下山。

Renovation day. I was not working on the mountain on this day but I was able to go and watch the renovations in between giving classes.

“伊凡亭”装修日。今天不是我在山上工作的时间，但在课间休息时我能抽空去山上看“伊凡亭”的装修。

As the 'Ivan Pavilion' nears the completion of its renovation, the architect poses in front.

"伊凡亭"的设计者于即将修葺一新的"伊凡亭"之前。

'Success'? I don't know if success can be measured by media interest but there has been a lot. I realise that as a foreigner I may have accomplished something that no native Chinese person could do. This has made me realise that I have a huge responsibility that I must fulfill if I am going to be true to the reputation that the media have given me.

“成功”？我不清楚成功是否可以通过媒体兴趣来计算，但是确实有很多媒体对此感兴趣。作为一个外国人，我认识到我获得了本土中国人不能获得的东西。这让我认识到如果要对得起媒体赋予我的美名，我要履行巨大的责任。

Another interview. Not all interviews were made by city or national media. This one was made by Wenzhou Medical University and I never did find out what they used it for.

又一次采访。不是所有的采访都是由温州市的或国家的媒体进行的。这次采访是温州医科大学做的，我一直没有弄清楚他们用这个干什么去了。

10 People Who Moved the Heart of Wenzhou. In 2012 I was nominated and judged to be one of the 10 People Who Moved the Heart of Wenzhou. This honour is something that still fills me with wonder two years later.

“感动温州十大人物”。2012 年我被提名并成为“感动温州十大人物”之一。即使在两年后的今天，这份荣耀仍然使我惊奇不已。

On stage—live TV. I have never spent much time in television studios but here I am on live TV, receiving a bouquet of flowers from a girl who performed admirably.

舞台——电视直播。我不曾在电视演播室待过很长时间，但此时我在电视直播节目上，接受一个表现非常优秀的小女孩的鲜花。

On stage after the interview—live TV. I made a real mess of the interview but still managed a grimacing smile afterwards.

采访后的舞台——电视直播。我将采访弄得一团糟，但仍在努力微笑。

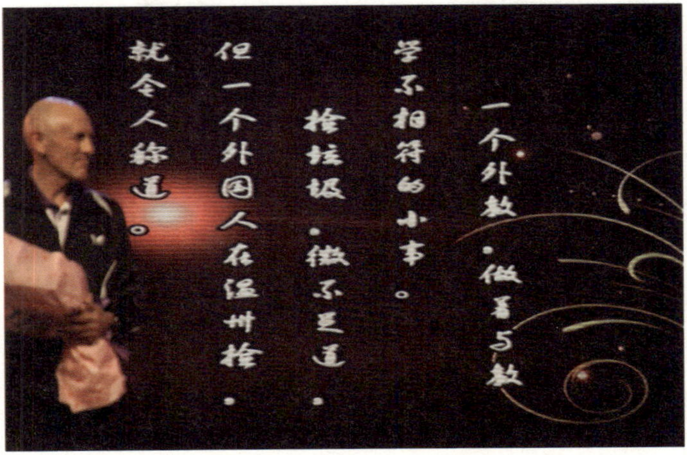

Listening to a dialogue about me. It seemed like the dialogue went on for an age as I stood on the stage listening and not understanding anything. If you are not Chinese and can't read the writing, I am assured that it says nice things about me.

聆听关于我的对话。我站在舞台上聆听，又听不懂任何东西，这使得这次对话好像进行了好几个世纪。虽然我既听不懂中文，又看不懂屏幕上的内容，但我可以确信他们是在赞扬我。

On stage with some famous people. Not everybody made it into this photo but there really were some fantastic people who received this award.

与名人同台。遗憾的是这次荣誉的获得者并不全在照片里，他们是真正了不起的人。

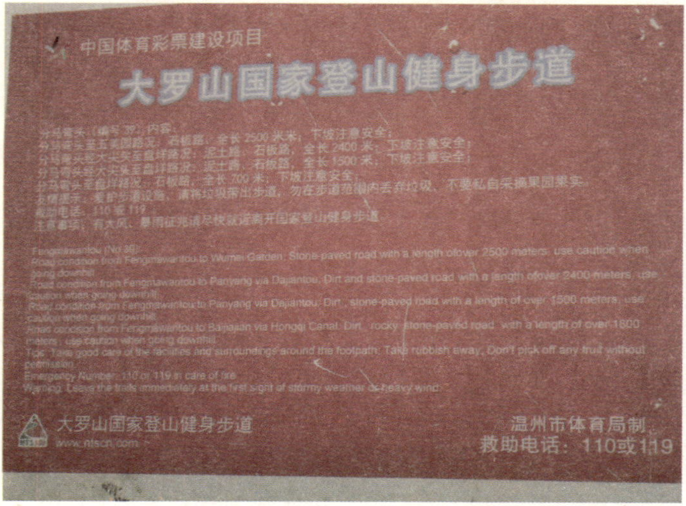

The government is trying to improve the tracks and obviously has some good long-term planning ideas. We even have some prohibition against dropping rubbish and other anti-social activities. This leadership needs grass roots support to work well.

当地政府正努力打造大罗山国家级健身步道，采取相关措施来禁止乱扔垃圾与其他不文明行为。政府需要来自民众的支持与合作，才能将这些措施贯彻下去，改善山道的环境。

TABLE OF CONTENTS

目　录

INTRODUCTION

Everybody has a story and this book is a part of my story. It is one part where I have done something worthwhile and I did it for long enough for a few people to notice. In the writing of the book I share some of the philosophies that I like. At times I hope you may think of this book as a success manual. That is rather presumptuous of me, because the only thing of any significance that I have done, became significant because I did it a few million times.

I hope you reach the end of this book because I hope you can find the couple of pearls of wisdom I stole from other people. I also hope that you may be affirmed in what you are doing, or that you will step up to whatever it is in life that beckons you, and that you will make a difference. I do not think that it takes the intellect of a rocket scientist to know that everybody makes a difference in our world. Every single person changes the environment in which they live. The big challenge to us all is: What type of difference are we making?

The difference I have made has been a small difference, but it has been measurable, and so here it is. This is a success story about improvement in the environment of a small part of China. China is a big country so the improvement here is like a small fish in a big ocean. If you add the rest of this planet that we call home then

简　介

每个人都有自己的故事，本册子就是我故事的一部分。这部分故事讲的是，我做了某件有益的事儿，而这件事儿我做了很长时间，结果引起了一些人的注意。在撰写本册子的过程中，我会分享自己喜欢的一些人生哲理。有些时候，我希望读者将本册子当作一本成功指南。我这样讲显得十分冒失，因为我所做过的唯一有些意义的事儿，就是因为我做了无数次才变得有意义的。

希望读者能“册”头“册”尾读完，因为我希望你从中能见到我窃自他人的几颗智慧之珠，我还希望你会坚信自己正在做的事情，或者坚信生命召唤你去面对的任何事情，并且希望你会做些改变。我认为，这个世界上，并非行家才知道每个人都可以做些改变。每个人都在改变自己居住的环境。我们大家面临的巨大挑战是：我们会做出什么样的改变？

我所做的改变不过是个极小的改变，但其意义非常深远。不妨读读这个励志故事，它讲述的是有关中国极小部分地区的环境改善经过。在中国这个泱泱大国里，这儿出现的改善仅是九牛一毛而已。如果你把这个我们称为家园的星球的其他地方也算进来，这“一毛”的类比就显得无足轻重了。该故事仍值得讲述，

the 'small fish' analogy fails to adequately describe the scale of insignificance. The story is still worth telling because there are lots of small stories like this. This message of hope amidst gloom needs to be told. People need to have hope for a better future because the situation for many, at the level they see it, is not good.

While living in China I have read in the *China Daily* numerous reports about various pollution problems. These problems are similar to those in other parts of the world and especially similar to those where development is taking place. I come from a place in the world where we have other people come just to experience the cleanest air on Earth. Here in Wenzhou the air is not so clean. I come from a place where the rivers are so clean it is normal to never see any rubbish in or along the waterways. Here in China I have not had that experience yet. During my last holiday to Australia I walked up Mount Wellington in Tasmania and I found two aluminium drink containers along the side of the road. That was all the rubbish I saw in a day's walking. I do not think you could replicate that experience anywhere in China but I could be wrong. It is possible here in Wenzhou to walk near the gardens that grow vegetables and strawberries without seeing plastic, paper, aluminium, and other assorted rubbish lying amongst the gardens. That experience however is not common. I have also experienced farmers overusing chemicals in orchards just as is described by Rachel Carson in her seminal work on the environment and unintended consequences, *Silent Spring*.

This latter example is just one of many that describe how our

是因为这类小故事是讲不完的。从灰暗环境中传递出的希望讯息有必要传述。人们需要对更好的未来抱有希望，因为就多数人所见层面而言，其处境还不够好。

在中国生活期间，我在《中国日报》上看到过无数污染问题的报道。类似问题跟世界上其他地区，尤其是发展中地区可谓大同小异。我来自一个允许其他地方的人同我们分享地球上最洁净空气的地方。温州这地方的空气有待提高。我家乡的河流非常清澈，无论是河道还是沿河两岸难觅垃圾的踪影。我在中国还不曾有这种经历。上次回澳大利亚度假期间，我爬上了塔斯马尼亚的威灵顿山，沿途见到两个铝制饮料罐，这便是我一天步行中见到的所有垃圾。以我之见，在中国任何地方都难以重现这种经历，但愿我错了。走在温州的蔬菜园、草莓园中，是可以不见废弃塑料、纸张、铝制品及其他分类垃圾踪影的，可惜那种经历并不常见。我曾见过果农在果园里过度使用化学药品，此举于环境有始料未及的后果，与蕾切尔·卡森在她开创性作品《寂静的春天》中的描述如出一辙。

在描述人与环境相互关系会如何影响生活质量的诸多事例中，

interaction with our environment affects our quality of life. The consequences of our actions are not always what we actually want. They too often show that we either do not care about the conditions we live in, that we have simply not thought about the consequences of our actions, or that we can't think ahead beyond what we are doing in the moment. In some cases it is simple ignorance. I believe it is this lack of awareness, lack of knowledge or lack of thinking about the consequences of our actions that is at the root of our environmental problems in the world. Of course short-sighted greed and self-interest are a part of that but those ideas are digressions to the main purpose of this book. If there is one indicator of our lack of thought for our quality of life and appreciation for the world we live in, I believe that our tendency to drop our rubbish wherever it is convenient to do so is the predominant example.

Wherever I have been in the world I have found people who will drop their litter because they do not want to carry it home, or carry it to the nearest bin. The scale of the problem varies but the fact that people consider it a problem is common to all. I would suggest that it is also more than just the superficial problem of being an eyesore or something leading to poor food and water quality. I suggest that it is an indicator of short-term thinking in regard to everything we do. The inability to work towards goals beyond even ten years is endemic in the government structures of most countries in our world. Many countries, including my own, have a three-year election cycle. This is re-enforced by the population's desires for quick fixes and fast rewards.

该例不过是沧海一粟而已。我们的种种行为带来的后果并非总如我们所愿。这些后果表明，要么是因为我们不在乎自己所生活的环境，要么是我们根本没有想到过自己行动所产生的后果，要么是我们的思维受限于目前我们正在做的事情。在某些情况下，这便是无知之举。我认为，出现世界范围内的环境污染问题，其根源就是我们对自己行为所产生的后果缺乏意识或者欠思考。当然，这些短视的贪婪和利己行为是环境污染问题的一部分原因，但这些想法并非本册子关注的主题。如果说有什么东西能表明我们对生活质量缺少思考，对自己生活的这个世界缺乏欣赏，我想随地乱扔垃圾便是一个有力的例证。

在这个世界上，但凡在所到之处，作者无不发现随地乱扔垃圾现象，因为人们不想把垃圾带回家，或者就近放入垃圾箱中。这一问题在程度上有所不同，但大家都认为这是一个问题。我想说的是，这不仅仅是有碍观瞻或者会导致劣质食物、水源污染这样的表面问题，就我们所做的一切而言，我认为这是一种短期思维的标志。世界上大多数国家的政府机构有一个通病，那就是其制定的工作目标不超过十年。包括澳大利亚在内的许多国家，其选举周期都是三年，而民众希望快速解决问题并得到回报的愿望进一步巩固了这个周期。

The focus of this book is China. In many ways the Chinese people with their strong government are able to work much closer to a ten-year planning cycle and even beyond that. This has obvious advantages. Those advantages are starting to be seen in care for the environment. All over the world environmental concerns have long been left behind in the rush for progress. Nowhere is this typified more than in the tendency for people to drop their rubbish wherever it becomes a burden to them.

This book was born out of anger and despair but the story goes on and it has become a story of hope, transformation, and beauty. I hope that regardless of whether you live in China or whether you have, or have not visited China, you will enjoy the story so far. I hope that it will give you hope and maybe even inspire you to further efforts in your part of the never-ending story.

本册子聚焦中国。中国人民在强有力政府的领导下，能够在诸多方面更紧密地朝着一个十年甚至更长期的规划而努力奋斗。该规划的种种优势开始显现于环保事业。就全球而言，人们因匆匆前进而长期忽视环境问题。最为典型的事例就是，一旦垃圾给人们带来不便，他们便表现出随地乱扔的倾向。

起初因愤怒与绝望而撰写的册子，随着故事的继续，变成了一个充满希望、变化和美丽的故事。迄今为止，不论你是否生活在中国，或者是否到过中国，我希望你会喜欢这个故事。我希望本册子能带给你希望，甚至可以激发你参与到这个没有结尾的故事中去。

CHAPTER 1

The Facts of This Story

As this book draws closer to a possible publication date in early 2015 you need to be aware that it has been written over a period of more than three years. There probably will not be a distinct ending to the story because right now I am working with Oujiang College as part of Wenzhou University and I do not recall ever working in such a pleasant environment, and so I am in no hurry to retire from what I am doing. Also the story is in no way constrained by my part in it. So there is no precise end point—it is hazy and indeterminate. Now, let me take you back to what may seem to be the logical start point.

In September 2011, I started working as an English teacher at Wenzhou Vocational and Technical College in the University Town of Wenzhou. This city is in Zhejiang Province in China. Within a couple of weeks of arriving I had walked up a few of the well trodden paths of Da Luo Mountain.

Less than a week after those first walks I started picking up rubbish on one section of these mountain pathways. At first I would go and pick rubbish for an hour or two several times a week but I quickly realised this task needed a long-term commitment and I would not be able to maintain that sort of effort for long. I then settled into a once-a-week routine on Sunday mornings.

第一章

故事中的真事

本册子日益临近2015年初出版之时，你可曾知其写作时间跨度竟有三年有余！书中记事很可能没有明确的结束时间，因为我目前仍任教于温州大学瓯江学院，我不曾有过如此舒心的工作环境，因此并不急于退离目前的岗位。书中记事绝不可因本人的参与而受到限制，因此也没有精确的结束点——只能模糊一些了。闲话少说，请随我回到故事的大致起始点。

2011年9月，我以一名英语教师的身份任职于温州大学城的温州职业技术学院。这个城市位于中国的浙江省。到达此地后的前几个星期里，我踏上大罗山上人们常走的那几条山道。

走完几条山道后不到一周，我便开始在这些山道的一部分区域捡拾垃圾。起初，我每星期上山好几次，每次工作一两个小时，但很快认识到这是一个需要长期坚持的工作。长期保持每周工作几次使我力不从心，于是决定每个星期天上午上山劳动一次。

It was not long and a few students started to accompany me. Mostly these were students from the Humanities Department of my college. I was working with the Mechanical Engineering Department but I got to know many other students by taking part in general college activities such as times when students would be focusing on practising English speaking and listening.

It was sometime after about six months of doing this that my little volunteer activity expanded dramatically. Someone amongst the administration of my college included it on the schedule of volunteer activities that the students could do. It was my understanding that students gained some sort of credit if they took part in some approved activities and so my regular Sunday morning excursions became one of these. This meant that my regular little group jumped from somewhere between two and six students to somewhere between six and forty-six. The overall effect of having all these extra workers was sometimes very good.

At about the same time that the volunteer workers increased in number, the media studies group from the college did a documentary of the activity and the point of that was to enter a national competition. I am told that they subsequently won that competition. This was the start of an ever-increasing media interest.

In general people were very interested in what I and the other volunteers were doing and were very positive in their attitudes towards us, and were also thankful for our efforts. This has led to awards and recognition for me personally and a general raising of awareness about pollution and specifically about the habit of dropping litter.

不久之后，几名学生开始陪着我上山劳动。这些学生大多来自我所在温州职业技术学院的人文系。我任教于机械工程系，但通过参与学院组织的一些集体活动，诸如学生集中练习听说英语的活动，我认识了其他系的很多学生。

约在捡拾垃圾六个月之后，我这个小小的志愿者活动规模开始急剧扩大。我所在学院的某位管理者还将此志愿者活动纳入学生志愿者工作的议事日程中。对此，我的理解是，如果学生参与一些经校方批准的活动，他们就获得某种程度的认可，因此我每个周日上午的惯例活动便成了其中的一部分。这就意味着，我小组的固定人数一下从 2 至 6 人陡增到 6 至 46 人。有了这些额外的工作者，其总体效果有时还真是不错。

差不多在志愿者人数增加的同时，来自学院的媒体研究小组录制了一个有关该活动的纪录片，旨在角逐一次全国性的比赛。有人告诉我，该小组在比赛中获得了名次。这便是媒体关注持续升温的开始。

总的说来，人们对我和志愿者们所做的事情非常感兴趣，对我们持有积极的态度，也很感激我们所做出的努力。这使得我个人获得了各种奖励和认可，也提高了大众对污染，特别是对乱扔垃圾这种不良习惯的认识。

What has been most significant has been a transformation in the amount of rubbish that is being dropped on the mountain and surrounding areas. There has also been an enormous government effort to clean waterways and improve rubbish collection routines. The government effort to clean the waterways of Zhejiang Province has been something with absolutely no connection to the efforts of me and all the people working with me, but it has helped to keep us motivated.

July 2013 saw me leave Wenzhou. I had some confidence that various groups of people would continue the work. This turned out to be the case. I hoped and I will continue to hope that the momentum will continue and grow because it is such a pity to see beautiful places spoiled by rubbish. It is such a pity to see people get used to living with rubbish around their streets and living areas all the time. I do not like to live like this and in general terms, nobody else does either. As I write this now in April 2014, I have returned to China and to my work on Da Luo Mountain. Whatever the future brings is immaterial to the fact that change has happened here. The change has been wide ranging, and it is gradually permeating all levels of the society.

最重要的是，山上及周边地区的垃圾数量出现了变化。为清理河道以及改善垃圾收集的日常工作，当地政府也做出了巨大的努力。政府清理浙江省河道的努力当然与我及和我一道工作的志愿者的努力全然没有关系，但这却成了我们工作的动力。

2013 年 7 月，我离开了温州。我相信各组织的成员会继续进行这项工作。事实也是如此。我曾希望，且会保持这份信念，这一势头会持续下去并壮大起来，因为如此多的美丽地方被垃圾破坏不能说不是一件憾事。看到人们对垃圾遍地的街道和社区里的生活习以为常，更是令人痛心。我不想这样生活，正常说来，没人想这样生活。2014 年 4 月撰写本册子时，我已回到了中国，也回到了大罗山开始工作。不管将来会发生什么，这对已然发生的变化并不重要。这一变化甚为广泛，渐渐影响社会各个层面。

CHAPTER 2

The Why of This Story

My actions in picking up rubbish have been a puzzle to many Chinese people and many people from other countries. They do not understand 'why' I might want to do such a thing and they definitely don't understand why I have kept doing it.

I believe it is important to examine this subject in detail because when we know 'why' we are doing something, then we can do it with focus. We can do it with certainty. We can do it with a right attitude, and we know where our power to do it and to keep doing it, comes from. If the reasons why we are doing something are clear enough in our head and those reasons are our reasons and are based solidly on our values and dreams, we can do anything. Facts, dogma, doctrines, culture, other people's expectations, historical precedents, these must all move aside and make way when the power of 'why' is fully functioning.

The reasons I started to pick up rubbish in Wenzhou on Da Luo Mountain are multi-faceted. It has always made me angry when people drop their litter anywhere but that is only the background reason why I started. It was however the reason that initially impelled me to action. I was angry that there was rubbish lying around in this beautiful place. I was angry that another

第二章

故事的起因

在不少中国人和外国人看来，我捡拾垃圾的行为一直是个谜。他们不明白我做这事的“起因”，当然更不明白我为什么还能一直坚持做下去。

我认为仔细审视这个话题很有必要，因为我们只有明白自己做某事的“起因”，才能一心一意去做，才能笃定地放手去做，才能以恰当的态度去做，也才会知道做这事以及一直做下去的力量源泉。如果我们对做某事的种种理由都能在头脑中形成足够清晰的认识，而这些理由刚好是我们所需要的，且与价值观和梦想紧密相连，那任何事情就会一做就成。当“起因”的力量得以充分发挥时，事实、信条、学说、文化、他人的期望及历史上的先例等都必须靠边让位。

我在温州大罗山开始捡垃圾，其原因是多方面的。人们随处扔垃圾总是让我愤怒，但那只是我开始此项工作的背景原因，然而，最初促使我行动的正是这个原因。我愤怒，是因为这美丽的地

person would think nothing of simply dropping their litter. I was angry that there was litter interrupting the line of my sight towards this wonderful expression of the natural environment.

The anger I felt was what got me started but I must immediately digress. I personally don't believe that anyone can make me angry unless I give them the license to do so. I have to abrogate my personal choice of feelings to someone else for this to happen. Normally I will not let that happen. I try to remain conscious of the fact that I create my own feelings and what is happening around me has little to do with what I create in this domain. It is my choice as to how I perceive these things and the results of that choice dictate what I may or may not be feeling.

After a lot of thought and procrastination I decided to choose to be angry about anybody making a mess in and on my home. This planet is my home (for now) and I like my home. I don't care if there are millions of other universes with exactly the same or similar Earths in them. This is one of the ideas that sit in the back of my mind and therefore I know it also sits with some other people. The idea, based on sound mathematics and quantum mechanics, is that there are many universes. I don't think that matters. This one, right here, right now, is my home and it is beautiful until someone spews their pollution into the air, pours their filth into the water, or drops their rubbish carelessly. I choose to be angry when pollution happens in any way because someone (including myself) has unthinkingly or selfishly left any sort of filth in their wake. I know that larger-scale pollution causes much more devastation than simple litter,

方遍地都是垃圾；我愤怒，是因为有人不把乱扔垃圾当什么大事；我愤怒，是因为自然环境在我视线中的宏大展现全让垃圾给打断了。

正是因为愤怒我才开始捡拾垃圾的，但我得马上岔开一下话题。我认为没人可以让我愤怒，除非我允许他们这样做。发生这种事，我必须搁置个人对他人性情的偏好。通常情况下，我不会让因事废人的情况出现。我尽力保持清醒意识：是我本人在制造情绪，在我周围所发生的事与我自己的情绪没多少关联。如何看待这些事情是我个人的选择，而这一选择的结果决定我将会产生怎样的情感。

经过长时间的思考以及久拖未决，我最终决定对把我家里家外搞得一团糟的人表示愤怒。(目前)这个星球就是我家，我爱我家。哪怕存在着亿万个宇宙，这些宇宙中存在着与我们地球一模一样或类似的星球，我也不在乎。这是我头脑中冒出的想法之一，我知道其他人也持有这一想法。根据严密的数学和量子力学理论，有很多个宇宙存在。我认为这没多大关系。此时此地的这个星球是我的家园，若没人污染空气、往水中倾倒脏物、随地乱扔垃圾，那它就是个美丽的地方。当人们(包括我自己)不假思索或者自私地随地乱扔垃圾而致使其遭到污染时，我选择了愤怒。我明白大范围的污染比小范围的垃圾会导致更严重的破坏，但我的观点是，这一切都始于人们随意丢弃一个汽水瓶，或者乱扔一个烟头。

but my thinking is that it all starts from the careless dropping of the soda can, or the flicking away of the cigarette butt. In different ways, pollution of any sort starts in our minds. All pollution starts with the lack of awareness that is epitomised by various actions or inactions. Correcting these behaviours, and the thinking or lack of thinking behind them, is where this little endeavour has started.

Why am I the way I am in regards to this sensitivity about rubbish? To answer this question would mean I would have to tell you my life story and that is not what this book is about. My Mum and Dad didn't like untidiness in any form but they were pedantic about visiting a place and leaving it tidier than it was when they arrived. Way back in the days when there were infinite places where you could camp freely and have a campfire in Australia, they had this attitude. They never left any rubbish behind and if a place had rubbish when they arrived, it had none, or at least a lot less when they left the place.

I remember my disgust one day when I discovered a soda can in the bush in the state of Victoria in Australia. In those days I liked drinking soda, but to find a can way out in wild bush country was very upsetting to me. I was in the army at the time and I was doing some extra personal training to improve my endurance and map-reading abilities. I had travelled a long way into state forests and parked my four wheel drive vehicle a long way from any made road. I was about half a day's walk from my vehicle and was climbing the side of a steep hill and struggling through thick bush. I remember thinking that there would

从不同的方式看，任何种类的污染都始于我们的头脑，所有污染都始于意识的缺乏，具体表现为各种行为或不作为上。纠正这些行为，找出其背后的观点或者意识的缺失，正是我此项工作开始的原因。

在对垃圾的敏感态度上，我为什么会依然故我呢？要回答这一问题，我得讲一个自己的人生故事，不过这与本册子无甚关联。我的父母亲不喜欢任何形式的脏乱，因此每当他们离开某到访之地时，总会让它比他们到达时更整洁。澳大利亚曾有数不清的地方可以安营扎寨、点燃篝火，远在那时，父母亲就已经形成这一习惯。他们从不随地乱扔垃圾。如果他们在到达某地时遇上垃圾，那在他们离开时，那个地方就没有垃圾，至少会少很多。

有一天，我在澳大利亚维多利亚省的灌木丛中发现了一个汽水罐，当时那种厌恶之情至今仍记忆犹新。尽管我当时喜欢喝汽水，现在仍然喜欢喝，但是在野生灌木丛深处发现的那个汽水罐让我心烦意乱至极。我当时在部队服役，正接受某项旨在提高我耐力和地图阅读能力的额外个人训练。我驱车很长一段距离，已经进入了密林丛中，将四轮驱动车停在远离道路的地方。我下车后走了半天的路，攀爬上了陡峭的山崖，在密林中穿行。记得我当时

have to be a possibility I was the first European Australian person to tread this part of our planet. At about that moment my foot slipped on a rock and disturbed the soda can. It was a very disappointing experience.

In a small city in southern China where I used to be until mid 2011, the rubbish dropping culture is pandemic in scale. Apart from further development of personal obsessive habits, I DID NOTHING about it there. I felt the anger rising within me and poisoning my physical and mental systems. I applied various techniques I have learnt to deal with those feelings, but to no avail. I could have made a real difference there but I DID NOTHING. I moved to Wenzhou in Zhejiang Province and things here were so much better. I saw people often put rubbish in bins and the place generally looked much better.

Here in Wenzhou I had mountains to walk up and it is there that I finally moved to action because it made me cry to see such a beautiful place spoilt. I decided on a particular track and I started clearing it of rubbish. I knew I could not fight all Chinese people, which disappointed the warrior within me, but I knew enough about the power of an individual and the power of the non-violent philosophy from the Gandhi tradition that I knew I could do something. Please don't think for one moment that I am being altruistic here. I deal with that subject in Chapter 4. This is all about me. My motivation in all facets was and remains selfish. Let's look first at the aspect of my anger. If I don't take action to address my anger then I will become a bitter, twisted,

还想当然地认为，我极有可能是踏上地球上该地区的第一位欧洲裔澳大利亚人。差不多就在那时候，我的脚在一块石头上滑了一下，碰到了那个汽水罐。那是一次令人非常失望的经历。

我曾在中国南方某小城待到2011年年中，乱扔垃圾的风气相当普遍。我个人的强迫性习惯虽有了进一步的深化，但在那里**我没有采取任何行动。**我心中的那种愤怒渐渐升腾起来，并影响着我的身心健康。我采用了各种各样学来的技巧来控制这些情绪，但全都徒劳。我本应该在那里采取行动做些实质性的改变，但是**我没有采取任何行动。**我迁往了浙江省的温州市，这里情况要好得多。我常常看到人们将垃圾放到垃圾箱里，这地方总的看来好得多。

在温州，我有山可爬。正是在这里我才最终采取了行动，因为我不忍心看到这般美丽的地方受到破坏。我确定了一条固定的山道，并开始清除那里的垃圾。我明白我无法跟所有的中国人抗衡，这一点使我内心的斗士精神受挫，不过，我对个人的力量和源于甘地式传统的非暴力哲学的力量有所了解，因此知道自己是可以有所作为的。千万别以此认为我是个利他主义者。这一话题我将在第四章中谈及。一切都是为了我自己。我在各方面的动机过去是，现在仍然是自私的。首先我们一起来审视一下我的愤怒情绪。

shrunken person. I will die earlier than necessary and I will spend more time being ill. I want your understanding to be absolutely clear about my motivation. I know the damage that anger can do to a person from the inside out.

Another facet of my motivation is my enjoyment of living on our beautiful planet. I could try to explain just how beautiful our small world is but I think that has already been done by our poets, photographers, musicians, and writers. I believe that each and every one of us knows that our world is intrinsically beautiful, even when we become aware of some of the ugliness we have to endure. When I see any place of beauty spoiled by any type of pollution, my enjoyment is lessened. I don't like that and I am quite sure that most people feel the same. The challenge may be that some people are not aware of their feelings or of how simple it is to keep a place beautiful or at least how simple it is to not have rubbish littering our environment.

Everything we do affects everywhere and everybody. This idea or concept is supported, at least to some degree by science, but let me give a simple example that is directly applicable to the subject of litter and my anger or enjoyment of our world.

One of my greatest joys in life is to run on the beautiful beaches of Australia. Unfortunately rubbish washes up on these beaches and it is brought there by the currents, tides, and waves from places all over the world. Years ago I remember reading a study done on litter. The results of the study proved that a small percentage of all litter that is dropped always ends up in the oceans. A small percentage of that litter ends up in my

如果我不采取任何行动来控制我的愤怒情绪，我将会变成一个厌世、扭曲、沉沦的人。我会疾病缠身，甚至过早死去。我想让各位真正明白我的动机。我明白愤怒情绪会从里到外给人带来伤害。

我还有一个动机，那就是我很享受生活在我们这个美丽的星球上。我本想将这个小星球的美丽诠释一番，但是诗人、摄影家、音乐家、作家等早已先我而行。即便我们意识到自己得不时忍受一些丑陋的事情，但我们这个世界在本质上是美丽无比的，我相信我们每个人都清楚这一点。当我看到一个美丽的地方遭到任何形式的污染，我的乐趣就会减少一些。我不喜欢那种感觉，我相信大多数人都有同感。问题是有些人没有意识到他们的情绪，或者没有意识到要保持一个地方的美丽是简单易行的，或者至少没有认识到不让垃圾影响环境是件非常简单的事。

我们所做的每件事影响着所有地方、所有人。这种想法或观念至少有一定的科学依据。请允许我举个简单例子，而该例直接针对垃圾与我的愤怒这一主题，或者我对世界的喜爱之情。

我人生中的最大乐趣之一，就是在澳大利亚那些美丽的沙滩上跑步。不幸的是，湾流、海潮和海浪将世界各地的垃圾冲到那些美丽的沙滩上。我记得多年前读过一篇有关垃圾的研究报告。该报告的研究结果显示，随地乱扔的任何一点垃圾终将会出现在各大洋中。这其中的一小部分垃圾于是就出现在澳大利亚我跑过的

pathway on the beaches of Australia. If I pick up a few million pieces of rubbish from a mountain in China, there may be fewer pieces washed up on those same beaches. I challenge you to find a flaw in my logic but I do ask you not to bamboozle me with talking about ocean currents. I have some understanding about the complexity of that subject.

A few years ago I was appalled to find out about the great areas in our oceans where rubbish has accumulated. Due to the general movements of the ocean currents there are now huge gyres of marine debris such as the Pacific Garbage Patch in oceans all over the world. If you have never heard of these then I challenge you to do ten minutes research on the Internet. If this problem keeps growing, and there is every indication that it will if something isn't done, then the prognosis for the health of our oceans on a global scale is poor. This was and is another bit of information that keeps me doing what I do. My hope is that if you are not already doing your bit for our environment, our future and the future of the human species, by the time you finish reading this book, you will.

Changing minds and patterns of behaviour doesn't usually happen very quickly. It does happen though. I am a part of change and hopefully that is change for the better. Everybody makes a difference wherever they are and whatever they do. The only question is: What sort of a difference will that be? I asked myself, 'What can I do?' The answer was that I could do the simplest thing possible and that was to not drop rubbish. I could also go one step further and pick it up. After that I could

海滩上。如果我在中国捡起亿万片垃圾，也许冲到那些沙滩上的垃圾就会减少几片。我恳请诸位找出我逻辑中的漏洞，但请不要用那些关于洋流的复杂话题来搪塞我。我对这一问题的复杂性多少知道一些。

几年前，我惊奇地了解到各大洋中那些聚集着垃圾的大片区域。由于洋流的运动，世界上各大洋都有巨大的海洋垃圾带，如太平洋垃圾带。如果你从没听说过这个信息，那就请你在网上做10分钟的调查。如果这个问题不断恶化，有确凿的证据表明就是那样的，若不采取行动，可以预测，全球范围内的海洋健康状况将会非常糟糕。促使我始终如一从事此项工作的另一项事实也就在此。要是你还没有为我们的环境、我们的未来，乃至整个人类的未来做出自己应尽的努力，我希望你读完本册子后就行动起来。

观念和行为模式尽管会改变，但通常不会轻而易举地改变。我就是变化的一部分，可喜的是，这是朝着好的方向在改变。不论身在何处，从事何种工作，每个人都在做出改变。唯一的问题是：那是怎样的一种改变？我曾经问过自己：“我能做些什么？”答案是我可以做再简单不过的事，那就是不乱扔垃圾。我还可以再向前

keep picking it up and just keep doing the same thing every week. It was simple. I could do that.

My leadership abilities are limited. My intellectual abilities are limited. My emotional abilities are limited. With all my limitations though, there are some things I can do and because I want to make a positive difference, I have decided I will do those things. The funny thing about all this is that I have discovered that in spite of having many limitations I am powerful beyond measure. I am so excited by that idea that sometimes I feel like I am bursting with motivation to find new ways to do something with that power. It also means that each and every one of us is gifted with our own unique combinations of power. It is unfortunate that many people have no idea just how powerful they are.

Why have I done what I have done in this rubbishy adventure? Why will I continue to do the same thing and a whole lot more? Now you know the answer—because I can, and it is good.

迈出一步，那就是捡起垃圾。之后我可以继续捡拾垃圾，并使之成为每个星期的习惯。这很简单，我是做得到的。

我的领导才能有限，我的智力有限，我的感知能力也有限。尽管受到诸多限制，但有一些事情我还是可以做的，因为我想做出积极的改变，于是就决定去做那些事情。有趣的是，我发现自己尽管受到诸多限制，却强大无比。这个想法让我非常振奋，有时我觉得自己可以用那种力量去发现新的做事方法。这也意味着我们每个人都有自己独特的力量。不幸的是，很多人却不知自己到底有多么强大。

在这场捡拾垃圾的“历险”中，我为什么做了这些事情？我为什么会坚持做同样的或者更多类似的事情？现在，你应该知道答案了——因为我能做到，因为这是有益的。

CHAPTER 3

Early Days

I have always loved mountains and since I first started running in about 1971, I have always loved running up mountains. When I was first taken to my accommodation room in my college and I saw the mountains close by I was delighted. Within a few days a student helped me to find some ways into the mountain area that is Da Luo Mountain. The beauty of the area was obvious and I had been looking forward to running regularly there but I was appalled at the rubbish littering the pathways wherever I went.

The anger in me about the rubbish was something that I did not like and I knew that I had to do something about it. Analysing my anger and having an intellectual understanding of it was not enough. I needed to take some sort of action quickly and so I decided to work on one area of the mountain path and to pick up the rubbish.

It is interesting now to look back at my feelings of the time because initially I was embarrassed to be seen picking up rubbish. I would go to the mountain at times when I knew there were few people there. Maybe I thought that people would laugh or poke fun at the weird foreigner doing menial work. Maybe I thought that picking up rubbish was a job for people who were not very

第三章

活动的初期

我一直热爱大山。自从1971年第一次开始跑步时，我一直喜欢沿着山道往上跑。当第一次被带到我在学院宿舍的房间时，一见近在咫尺的群山我便高兴不已。没过几天，一名学生带我找到了通向山区，即大罗山的一些道路。山区风景自然美不胜收，我期望定期去那里跑步。可是我所经之山道皆满布垃圾，这让我惊愕不已。

我不喜欢由垃圾引起的愤怒情绪，因此明白必须得采取相应的措施。仅分析和理性认识自己的愤怒情绪还远远不够。我得迅速采取某种行动，因此决定在山道的一个区域开展工作：捡拾垃圾。

现在，回顾起我当时的心情还挺有趣的，因为起初让人见我捡拾垃圾很是尴尬。我有时会趁山上人少的时候才去那儿（捡垃圾）。也许我认为人们会嘲笑我是个做卑微工作的古怪外国人；也许我认为捡垃圾是那些并不聪明的人才干的活，我自己去捡拾垃

clever and by doing the job myself I was admitting that I was not very smart. Maybe it was both of these reasons but it is worth admitting to the world that I was embarrassed to be seen picking up rubbish. I can also remember being embarrassed by my own father picking up rubbish but there I was doing the same thing and I was still embarrassed.

Somehow I managed to process through that thinking and to replace it with a warrior's type of attitude where I was willing to stand up and to be seen. I became willing to stand up and to be a good worker, servant, and an example. This was not an easy change to make and I have seen many people come to do this work but then be unable to make a similar change. The embarrassment stayed with them and so they could not keep doing it even though intellectually they could see it was a good thing to do.

Why? It is worth examining this question because this very reason will stop many people from bending down and picking up that unsightly bit of trash. I guess there is more than one reason so let's start. I think pride is the best place. Isn't it a powerful emotion? I was afraid of what people might think of me and I guess that other people are the same. I was afraid that they would think I was a loser, the equivalent of the old street person who had always been on the losing side and on top of that he smelled very bad. I was afraid that people would think I am some kind of weirdo. I wondered what kind of weirdo they would think I was. I guess that other people are also afraid of what their friends or even a stranger might be thinking about them.

圾就等于承认了我不够聪明；也许两个原因都有。总之，我得向世人坦白，被人看见捡拾垃圾时是尴尬的。我仍然记得，见我父亲捡拾垃圾让我感到尴尬，而此时我也在做同一件事情，我仍然尴尬不已。

好在我最终克服了这种想法，用一种斗士的态度取而代之。我愿意站出来让人观看。我愿意站出来，愿意成为一名优秀的工作者、仆人、榜样。做出这样的改变并不容易。我看到很多人来做这个工作，却不能做出类似的改变。尴尬的感觉伴随着他们，因此他们不能继续做下去，虽然理智告诉他们这是一件有意义的事情。

为什么这么说呢？这是一个值得审视的问题，因为它正是阻止人们弯不下腰去捡起那点不堪入目的垃圾的原因。我想原因还不止一个，因此不妨说说。我认为骄傲才是最重要的原因。难道骄傲不是一种强大的情感吗？我担心别人会怎样看待我，我想其他人也会有这样的顾虑。我担心人们会把我看成是一个失败者，或者说是一个年迈的街头流浪汉。这种人总是失败，最糟糕的是，他体臭难闻。我担心人们会把我当作怪胎。我不知道他们到底会把我当作哪一类怪胎。我想别人也会在意朋友甚至陌生人对他们的看法。

None of this stands up to logical argument. Although I have had times when I was on the losing side of life, generally I have done pretty well. Many people have a similar story to tell. I don't smell bad although I know that I do smell bad if I don't shower or clean myself every day. Certainly I smell bad after a few hours of picking up rubbish. If some people think I am weird then that is surely their concern and I shouldn't try to change their minds. I think it is really weird for people to just drop their rubbish anywhere without thinking about the consequences for their standard of living. Do those people give one single thought to what I might think about them? No, they don't. Would it be a good thing if perhaps my weirdness started them thinking a bit? Yes, and that is a part of what this project is all about.

The next thing that was important for me was more to do with the attitude with which I worked at the task. When people are discussing leadership, it is common knowledge that if you get into a confrontational state with somebody, any change on their part will be reluctant and will probably only ever be given begrudgingly. Also any change that they exhibit is likely to be short-lived. If one can persuade for change by leading by example, then that change is more likely to be complete. This is even more so if the example is given with the complimentary, gracious attitude of a willing servant. Somehow in those early days I slipped into this attitude but it was not always easy to maintain as time went on. When I slowly came to realise how famous I had become in Wenzhou was one of those times. There

所有这些都经不起逻辑推理。虽然生活中的我也有失败的时候，但总的说来，我做得还是相当不错的。很多人都有相似的经历可讲。我没有体臭，我明白如果我不每天洗澡，我的身体确实会发出异味。捡拾几个小时的垃圾之后，我身上当然有异味。如果有些人因此认为我是怪人，那是他们的事，我不必去改变他们的想法。那些不顾垃圾影响生活水平而随地乱扔的人，他们才是真正的怪人。他们考虑过我对他们的看法吗？没有，他们没有。要是我的怪异行动能让他们开始对此略有思考，这难道不是件好事吗？是的，这正是整个项目的一部分。

对我来说，下一个重要的事情与我从事此项工作的态度更有关系。当人们讨论领导才能时，人所共知的是，如果你跟他人处于对立状态，那么双方做出的任何改变都是勉强的，甚至是情非所愿的。再有就是，他们所表现出来的任何变化都可能是短期的。如果有人能带头做出表率来说服他人，那种改变才有可能是彻底的。如果像一名顺从的仆人那样用亲切和赞赏的态度做出这种表率，这种改变则会更加彻底。不知何故，早些时候，我渐渐养成了这种态度，但是随着时间的流逝，要保持这种态度并非易事。当我渐渐

is no doubt in my mind though, that this is the best possible attitude to develop, to maintain, and to cherish. It is the antidote to the big ego. It is the antidote to feelings of superiority. It is the antidote to any 'holier than thou' type of attitudes. On the other side of the coin it still allows for pride in a job well done. It still allows for courage, determination, persistence and for big dreams.

Within a couple of weeks of starting my project to clear one kilometre of track, a few students had started coming with me and I had decided that this would happen at the same time every week and we would leave from the same place. The reasons for this were various.

'Habit force' was the first reason. I have done a little bit of study about the power of habits in our lives so I knew that if I could create the habit then a lot of benefits can flow from that. Once a habit is formed then it is quite easy to maintain and it can also make a difficult task easy. If a habit is chosen and implemented for good reasons then it has a close relationship to personal discipline. When you combine personal discipline with habit force, you then have something that can give you the freedom to succeed at projects which would otherwise be impossible. Habits in combination with personal discipline can overcome even our own feelings. Too often we allow our temporary feelings to get in the way of our long-term success.

The second reason that I set the same time and place to leave every week was to cut down the number of phone calls and texts that I would have to deal with on a weekly basis. Communicating

意识到自己在温州的那点知名度，我发现很难再保持这种态度了。毫无疑问，这是我要养成、保持和珍视的最佳态度。这是消除自负的良药；这是消除优越感的良药；这是消除“比你更圣洁”式态度的良药。从另一方面看，这种态度体现了因做好了工作而产生的自豪之情，这种态度体现了勇气、决心、毅力以及宏伟的梦想。

在着手清理一公里步道垃圾的几个星期里，一些学生开始同我一道上山，我决定将这固定在每星期的同一时间，我们可从同一地点出发。这样做有很多原因。

“习惯的力量”是首要原因。我做过一点关于生活中习惯的调查，明白自己要是能养成一种习惯，就会带来一系列好处。一旦习惯养成，就很容易保持，这也会让艰难的工作变得容易。当你有充足的理由选择并开始培养某种习惯时，你会发现培养这种习惯与个人自律有着密切关系。当你将自律与习惯的力量结合起来时，你就拥有了做某事并能成功的自由，否则是绝无可能的。与自律相结合的习惯甚至可以克服我们的情绪。可我们很多时候却让暂时的情绪妨碍自己长远的成功。

我定于每星期同一时间同一地点出发的第二个原因，是减少我每周必须处理的电话和短信量。对我来说，与中国人交流不总

in China has often not been easy for me, but I could make it easier by simply learning how to say the same thing every time and then to simply repeat it.

The third reason was to take away excuses for not coming with me and doing the work. When people understood that I was serious about the work and that I was reliable, then they always knew where I would be and what I would be doing. It didn't matter what the weather was going to be; I would be there. Sometimes in the early days Catherine and I would be the only people walking up the mountain in the rain and sometimes we would turn back after an hour or so, but we always went. Later when Catherine moved on to focus on her studies and then to work, it was Betty and I. I think Betty was to this point at least, the person who has been the most consistent for the longest time in sticking with me through times when we had a crowd of people with us to when it was just her and me.

In the early days I started by using plastic bags to put the rubbish in and plastic gloves to protect my hands and I kept doing this and supplying students with the same kit for many weeks. Obviously using plastics like this is not good for the environment so eventually I found good substitutes. Cotton gloves that I wash and continue to wash after every Sunday are excellent. Although they are of poor quality and often need a bit of repair work, they are of excellent value and actually protect our hands much better than plastic gloves. They do get very wet and smelly but they still do their job of protecting our hands from cuts and prickles. I found a hessian type of bag that we are able to use many times

是那么畅通，但通过学会每次说相同的事情，只是简单地重复一遍，我就能让交流顺畅许多。

第三个原因是排除那些不跟我去做事的人的种种借口。当人们明白我对此项工作是认真且靠得住的时候，他们就会知道我将会出现的位置以及我将要做的事情。不管天晴还是下雨，我都出现在那里。早期，有时只有我和凯瑟琳冒雨上山，有时我们会在一两个小时后返回，但我们总是坚持上山。之后，当凯瑟琳集中精力学习，后来参加工作后，是贝蒂（张玲燕）和我一同上山。至少到此为止，我认为贝蒂是坚持得最久的人，有时我们有一大群人，有时只有我和她。

早些时候，我用塑料袋来装垃圾，用塑料手套来保护双手，好几个星期以来，也为学生们提供塑料袋和塑料手套。显然，使用这样的塑料制品不利于环境保护，最终我找到了理想的替代品。棉手套非常管用，这种手套可以每周清洗后重复使用。虽然质量不佳，往往还要进行小修小补，但这东西很有价值，能比塑料手套更好地保护我们的双手。这种手套会在弄湿后发出难闻的气味，但

before they eventually fall apart. They also need to be washed at the end of each session.

The principle that is worth remembering about this is that it is better to take action than to simply think about doing something. Obviously some thought should happen before we leap into action sometimes but nothing happens without action. Once action or movement is happening it is a relatively simple thing to change course a little. Mistakes can be corrected easily when movement is happening. When nothing is happening, there are no mistakes and no progress.

Often in the past I have made the mistake of doing nothing and accomplishing nothing. Sometimes in the past when I did things I made mistakes. I need to remember that at least when I was doing something and making mistakes, I learnt something in the process. My real mistake was that it should have been a rare occasion to do nothing and that frequently I should have been doing something and therefore be able to make a lot more mistakes. I now know that the mistakes we made in our early days picking up rubbish were very worthwhile. I still make mistakes but now I simply try to learn from them quicker than I used to do.

它们仍可以保护我们的双手，使其免遭荆棘划伤。我找到了一种麻布口袋，可以反复使用多次，但每次使用后都得清洗一遍。

有个原则值得记住，那就是心动不如行动。显然，有时在投入行动之前我们是需要有所思考的，但是如果没有行动，一切都不会发生。一旦有行动，那么略微改变一下行动方向就相对简单多了。如果有行动，错误也容易纠正。如果没有行动，虽然不会犯错误，但却不会有进步。

我过去常常犯的就是无所作为而无所成就的错误。过去有时我做事就犯错误。我有必要记住的是，我至少是在做事的时候才犯错，可也从中学到了东西。我真正的过错是：无事可干的场合毕竟是少数，我本该经常做些事，才有机会犯更多的错误。我现在认识到，我们早些时候捡拾垃圾时所犯的错误是非常值得的。我现在仍然犯错误，但比过去更能快速地从中吸取教训。

CHAPTER 4

Progress and Strange Happenings

This chapter is a collection of bits I wrote at odd times along this journey. They may not make a lot of sense to you from a time line perspective but I can assure you that a lot of what has happened did not make sense to me at the time. Most of these bits are taken from letters I wrote to folks in Australia.

Part 1

For about 2 or is it 3 months now I have been going up Da Luo Mountain and picking up rubbish from a one kilometre stretch of track. In that time I have seen many little changes and some that are really not so small. Last week there were big, brand new environmentally green rubbish bins at all the rest points. Previously the only bins were these small bin type things with a smallish side opening and a larger opening at the bottom for the rubbish to spread out onto the ground. They must have been a good idea at some time to somebody but now they are just one of the silliest things I have ever seen.

As a consequence of these new bins there was far less rubbish to pick up this week so, another section just off the track was cleaned up effectively. There are still a couple more areas like this where quantities of rubbish have been thrown into the bush.

第四章

进步与奇事

本章由作者写作历程中的偶拾汇编而成。从时间这条线上来说，其中内容对读者大众或许没太大的意义，可我担保当时发生的很多事情，我当时看也没什么意义，现在看就不同了。这些偶作中的大部分摘自我写给澳大利亚友人的信件。

第一部分

我上大罗山捡拾一公里步道区域内的垃圾，到此为止已历时两三个月。其间我目睹了许多小变化，其中一些还真不小。上个星期，在沿途各憩息之处都配置有崭新的大型环保垃圾箱。以前仅有的那种垃圾桶容量偏小，边上留有小口，底部大开口，结果让垃圾散落得满地都是。这在当时对某些人来说准是不错的创意，但它现在却成了我见过的最傻气的东西之一。

由于有了这些新的垃圾箱，那一周要捡拾的垃圾明显有所减少，步道另一边区域也得到了有效的清理。山上还有好几处类似的区域，大量的垃圾被抛进了灌木丛中。人们不假思索地将这些

Without any obvious thought people have used the area as the local rubbish dump creating an eyesore that takes so much away from the beauty of the mountain and is a testimony to laziness and short-sighted thinking.

On other occasions I have seen other people who are also trying to clean up the rubbish and make the mountain a more beautiful place for all to enjoy. There was a couple who looked like they were very focused on cleaning a particular area. There was a group of about 20 people all dressed up in matching caps and carrying matching bags and picking up rubbish gadgets. They seemed to be more show than substance as there was a lot of rubbish left behind where they had been walking, but I must not be too hasty to judge them. They could easily have been heading to work in a different area. It was nice to see them anyway because it indicates awareness and lack of awareness is the real problem.

Part 2

Most of you would be aware I have taken on a task of picking up rubbish once a week for a few hours on a mountain path within walking distance from my home base here at the college where I work. This has been a consistent activity now for about 19 months or so except for when I was away on holidays. Someone here at the college organised a system of volunteers from the student body and most Sunday mornings I have a group of up to 30 students who join me. If it is raining or they have other activities then they don't come, so then it is just

区域当作垃圾倾倒场，这人为造成的碍眼之地极大地破坏了山中美景，也暴露出人们的惰性和短视。

在别的地方，我也见到其他人在努力清除垃圾，好让大罗山成为一个人见人爱的更美之地。有一对夫妇，他俩似乎在集中精力清理某一特定区域。有一个约 20 人的小组，他们带着清一色的帽子，手里拿着清一色的袋子和清一色的垃圾捡拾器。与其说他们是在做实事，倒不如说是在作秀，因为他们经过之处还留有不少垃圾。对于他们的行为，我不应该草率判断，他们有可能正在赶往某个不同的工作区域。不管怎么说，见到他们还是令人高兴的，因为这表明人们已经意识到了需要保护环境。而缺少环境意识恰恰是真正的问题所在。

第二部分

诸位大都已经知道，我每周在山道上捡拾数小时垃圾，那个地方从我任教的“总部”走路便可到达。除去外出度假的时间，这项活动我已持续了大约 19 个月之久。学院有人从学生会中组织起了一批志愿者，在大多数周日早上，有大约 30 人加入我的活动。

me or me and a couple of other students who are particularly disciplined with wanting to practise English with me or who are genuinely concerned about the environment.

Over time my face has become fairly well known and to cut a long story short, someone nominated me to be amongst a group of people who were chosen because they 'moved the heart of Wenzhou', —and so it happened.

There was an internet site that was made in preparation for the presentation night for the top 10 people who had 'moved the heart of Wenzhou' in the past year. I still don't know how or who nominated me for this thing but it seemed to be building its own momentum for some time leading up to this. It was certainly way outside any control I had. When I became aware that I had been nominated and there was a voting process it was all too late to make any difference. In spite of my lack of knowledge about this event and its processes, obviously I still was voted to be one of these 10 people.

On the night of the presentations I found I was sharing the stage with a very eclectic group. They included a girl who had done CPR on a bloke who had drowned and so she saved his life. There were a couple of police officers who had died in the performance of their duties. Their immediate family and work comrades were represented. There were a few groups of people who had done odd things. Some I never worked out what they did but one lot had started a kitchen to feed the starving poor.

如遇下雨或者有别的活动要参加，他们便暂停活动，就只有我一个人上山，有时几个特别想跟我练习英语或真正关心环境的学生也会和我一起坚持上山捡拾垃圾。

随着时间的推移，人们渐渐熟悉起我的面孔来。长话短说，有人提名我为“感动温州十大人物”的候选人，结果我还真就成了。

网上有去年“感动温州十大人物”颁奖之夜的报道。我怎样或是被谁给推荐上去的，我至今也不知道，可能是日积月累后水到渠成，但这显然已经超出了我的掌控范围。当我后来意识到已获提名，并且还要经历一个推举过程时，想要做出任何改变已为时已晚。尽管我对此活动及其程序所知甚少，但还是成为了“感动温州十大人物”。

颁奖的那天晚上，我发现自己与许多人共享一个舞台：有位女孩，她替一位溺水者做心肺复苏，最终挽救了他的生命；几名以身殉职的警官的直系家属和战友亲临现场；还有几组从事古怪行当的人；有些人做了什么我无从得知，但我知道有一组人为饥饿的穷人办了个厨房；有位天生腿脚不方便的人成了一名优秀的中学教师，后来又回到课堂学习别的知识；另有一名教师致力于贫困地区的教学；还有一名边求学边照顾多病祖父母多年的女孩；加上一名

There was a bloke born with brumby legs who had become a very good high school teacher and then was back studying for something else. Another teacher had dedicated himself to teaching in some poor area. A girl had looked after ailing grandparents all the time while getting herself to school and then there was one foreigner amongst them all who picked up rubbish. Amazing stuff! There were countless newspaper people and other media people at this 'live' TV ceremony but also my college here bussed in about 50 students from my department as well as their media bloke and a couple of teachers or whatever. I never really know what is going on.

Anyway, the next day it was back to the normal business of picking up rubbish on the mountain. There is now a bloke who is obviously paid to sweep the paths and keep the bins clear. I have no idea how often he goes but it is another thing that is making a difference. The biggest thing though is that there is a lot less litter being casually dropped. I personally believe that until people are willing to take care of this little thing, there will never be the political or general will in any society to take care of the bigger things. Bigger things such as air and water pollution and the different pollutants that impact on all the food that people have to eat. I learnt not long ago that when rich folk here become aware of how bad the food is, they buy imported food. In a country with a population of a billion people that becomes something pretty significant I think.

捡拾垃圾的外国人。多么有趣的组合！出席这场“直播”颁奖仪式的，不仅有不计其数的记者及其他媒体人士，还有学院用车送去的我所在分院与传媒系的约50名学生、几名老师等。我真不知道是咋回事。

尽管如此，我第二天仍然例行上山捡拾垃圾。目前，有个显然是受雇来清扫山道并清理垃圾箱的人。我不知道他多久上一次山，但这是正在发生的变化之一。最大的变化是随地乱扔的垃圾大大减少了。依我个人之见，任何社会，只有到人们愿意关注捡拾垃圾之类的小事之时，才会有政治意愿或者民众诉求去关注诸如空气污染、水源污染以及食品安全问题等更大的事。不久前，我了解到这里的有钱人已意识到食品的糟糕程度，他们购买进口食品。在一个拥有十亿之众的国家，我认为这是一个相当严重的问题。

Part 3

I am sorry this is very haphazard writing, but I seem to be a bit hurried these days. This thing took up a lot of my time and effort and so my administration fell a bit behind, but the biggest thing that is taking up my time is Ping Pong. Here I am in Ping Pong paradise and I think one of the best coaches that could ever be, thinks I am something special, so I am spending far too much time learning to play this crazy game. My excuse would be that you have to make hay when the sun is shining but really I am just having fun.

Part 4

I have been laid up from playing table tennis for a couple of days because I blew out my arm so I had to finish marking examination papers. I should have been doing this anyway as a priority way above Ping Pong so there you go. Several of the students' essays included bits about me. Most of these essays were of very poor quality. I think maybe these students were trying to appeal to my vanity. They picked one of my faults to a tee (vanity) but unfortunately for them it didn't help them get good marks. A couple of them suggested I was now famous throughout Wenzhou. After treating those comments with some derision I was then astounded to receive an e-mail from an ex-student from southern China and evidently I had made in onto their TV screens. Go figure that! I ask myself: What is this all about? Right now—I don't know. I simply don't understand it.

第三部分

很抱歉，这样写作太随意了一些，我这些日子来好像有些忙。这件事占去了我大量的时间和精力，以至于使我的本职工作有些滞后，而花去我时间最多的事就是乒乓球。我正身处乒乓球的天堂。我想，迄今为止最优秀的教练之一认为我有特殊才能，于是就花了过多的时间去练习这疯狂的球类。我给自己找的借口叫作“趁热打铁”，当然也只是在玩而已。

第四部分

因为胳膊受伤，我好几天都未能玩乒乓球了，因此只好把试卷批完。我理应先批阅试卷才是，而不是玩什么乒乓球，让你说着了。几个学生的作文都提到了我，大多数作文写得很差。我想这些学生大概在迎合我的虚荣心。他们奉承我，但不幸的是，这没能帮助他们取得好成绩。有几个学生还认为我现在红遍了整个温州。我带着几分嘲弄的心情处理完这些作文，还惊讶地收到了一封来自华南的电子邮件，是从前的一名学生发来的，显然是因为我上了他们当地的电视屏幕。真没想到！我问我自己：这到底是怎么一回事？到目前为止，我啥都不知道，我真感到莫名其妙。

Part 5

May 5,2013

I would be of the opinion that most transformations need careful nurturing and continued growth, otherwise regression is certain. Anyone who has been through any sort of recovery process would agree. There comes a point, sooner or later, that the transformation may seem to be self-sustaining, but there is still that positive input that is happening from some source or another.

Today as usual, I led my troupe of newbie volunteers up the mountain to pick up rubbish, but today was different in two ways. The first was that we had not been up the week before so it had been two weeks since we cleared our two kilometres of track. The second reason today was different, and was in fact that two weeks ago I sprouted on about this huge transformation that I could see. I can happily confirm that the transformation is being maintained by forces outside of my/our control. It was a small bag day and with a handful of the volunteers really working well we extended our territory into some difficult places where we had never been before.

Part 6

May 24, 2013

I see more and more little things that have improved without our direct influence and they may be improving without having any influence from me or my team at all. I will never know but it can still be a secret delight that maybe, just maybe, it is.

第五部分

2013 年 5 月 5 日。我同意这一观点，那就是大多数变化需要细心呵护且持续发展，否则倒退是确定无疑的。任何经历过康复过程的人都会认可这一点。还有一点，那就是变化迟早可实现自我持续，但仍不断需要来自某种渠道的正面引导。

今天，我一如既往地带了一队新的志愿者上山捡垃圾，但今天却有两方面的不同。第一个不同是，上个星期我们没有上山，因此已有两个星期没有清理那两公里的山道；第二个有别于往常的是，两个星期以前我大讲特讲我所看到的这一巨大变化。我欣喜地证实，确实有我（或我们）之外的种种力量正维持着这一转变。多亏了几个志愿者的认真努力，我们得以将清理的地界扩展到了我们以前不曾涉足的困难区域。

第六部分

2013 年 5 月 24 日。我发现，越来越多的小事情无需我们的直接影响就出现了改善的迹象。也许它们并不需要我和我的团队的任何影响都会发生改善，这我永远无从得知，但是让人不禁暗喜的是，这变化也许——只是也许——归功于我们的努力。

Part 7

Last week we extended our territory so that we began work before we actually got to the foot of the mountain. The pathways, roadways and tracks we take are really ghastly in places. The littering has been happening for years and it keeps happening. Occasionally a street cleaner might pass through but they fight a losing battle. Anyway we started in an alleyway outside a middle school about 200 metres before the mountain. After this little alleyway there is a section of normal suburban street, and we kept at our task there. The reaction of the locals was precious. There was obvious delight from some of them and the usual wonderment from others. The weather on this day was one of the first hot and humid days of the season. My group of volunteers started out at about 20 but by the time we had been working for half an hour or so the number had dropped to 6 and only 4 of them were working. But what great workers they were. They were all girls and they were as brave and daring as one could possibly hope for. Under normal situations, the Chinese people and particularly these students are very hesitant to go off the track and chase the rubbish into the prickles and onto the treacherous footing on the fairly steep side of the mountain. A couple of these girls were heading into it while I was still hesitating and working out a way to do it. I was mightily impressed. They were real warriors. I hope I see them again but the likelihood is small. I seem to get a brand new group of volunteers every week. Last week I wrote a 'THANK YOU' internal memorandum to the unknown administrator

第七部分

上个星期，我们扩展了自己的清理区域，因此，事实上，不等到达山脚，我们便开始工作起来。我们经过的人行道、小道、车道的有些地方脏得可怕。乱扔垃圾现象多年以来一直存在，并且还在继续。偶尔有清洁工人从面前经过，但他们在打一场赢不了的仗。不管怎么说，我们在离山脚还有两百米的一所中学外的小巷开始了清理工作。在这个小巷之外，是一段郊区的正规街道，我们将在那儿继续干下去。当地人的反应值得重视：一些人明显喜形于色；另一些人则表现出其常有的不解状。当天是初夏特有的那种闷热潮湿的天气。刚开始时，我那支志愿者队伍有 20 人左右；干了半小时后，只剩下 6 人，其中干活的只有 4 人，都是女孩。她们还真是些了不起的劳动者，表现出超乎想象的勇敢大胆。通常情况下，去步行道外的荆棘丛或者陡峭且无法立脚的地方捡拾垃圾，中国人，尤其是那些学生都会踌躇不前。正当我为寻求办法而犹豫之际，几个女孩却走了进去。这给我留下了极为深刻的印象。她们是真正的勇士。我希望能再见到她们，但可能性很小。每周似乎都有新的志愿者加入。上个星期，我给那位不知名的管理者写了一份“致谢”的内部便函，是他组织了那些志愿者小组加入我的活

who organised that the volunteer groups would join me on the schedule of works that they do as part of their system of getting points toward their degrees, or opportunities for scholarships, or whatever it is. You will note that I still haven't worked out what their system is.

Will this transformation be maintained after I leave? Of course I hope that it will but it is outside my control. I encourage students but I am afraid that without a leader saying to the group every Sunday morning 'Let's go!' it may fade away. I even tell the obvious leaders amongst the students that that is about all they have to do because I know they lack the confidence of us brazen old folk. Ha ha. It will be what it will be.

Part 8

One of my problems in doing this work has been walking to the site. On the way there are a couple of streams/rivers and they were full of rubbish. There was still some place for water to sit or haphazardly flow through the rubbish but it was truly disgusting. In the last few weeks these rivers have been transformed. Workers have removed the bulk of rubbish and so transformed the sites into recognisable waterways. I can't help noticing that there are small amounts of rubbish appearing again but on the largest river there is now a sign with the names and phone numbers of people responsible for the waterway. Presumably one could phone and report if dumping or other poor use of the river was happening. I have wondered if I and my group had any

动，并在工作日程中规定这些活动能够让学生在获取学位、奖学金或别的什么评选中得到加分。其实我依然不明白他们的操作流程。

这种转变在我离开后还会保持下去吗？我当然希望它能够保持下去，但这不在我的掌控范围内。我鼓励学生继续这项活动，但是我担心没有领头人每个周日早上发出“出发”的指令，这一转变会慢慢消失殆尽。我甚至告诉那些一看便知是领头人的学生必须这样做，因为我知道他们缺乏我们这些脸皮厚的老家伙的自信。哈哈，将来还是顺其自然吧。

第八部分

从事这项活动，我所面临的问题之一是要步行去工作场地。途中有几条满是垃圾的小河。一些地方的水滞留在垃圾的包围中，或者流过杂乱的垃圾，那景象看起来实在令人恶心。近几周来，这些小河发生了转变。工人们清走了巨大的垃圾堆，将这些臭水沟变成了一眼便可认出的水道。少量垃圾的再次出现，不禁引起了我的注意。不过，在那条最大的河边竖着一块牌子，上面留有河道负责人的名字和电话。我猜想，要是出现倾倒垃圾或者滥用水域的行为，人们可以打电话举报。我一直纳闷该不是我或者我

influence on this happening. I have also wondered if it will be maintained. This is certainly something way outside of my control but I know what I can control and that is consistency of effort doing what I do.

Part 9

I have recently found out from a reporter that we had nothing to do with the rivers being cleaned. This was a Zhejiang Province wide initiative by the government. I'm sure it costs them a lot of money. Recently I noticed that people are growing some type of vegetable or something in containers in the waters of these rivers. The water still does not look very good but it is a mighty lot better than it was.

Part 10

Talking about water, I had a reporter ask me about how much cleaner the water was in a reservoir up in the mountains where we have done a lot of work over these years. The sides of the dam wall and along the water used to be littered with rubbish but now there is only the occasional new bit of litter to be picked up. Actually there is a favourite picnic/BBQ spot as well where people often leave a pile of rubbish for us. That is always disappointing. We will win the battle one day. Anyway, my point is that the reporter was sort of trying to get me to claim responsibility for this clean reservoir and I did not. Why? There were a lot of students and other people who picked up a

们小组对此产生了些许影响。我也一直纳闷这一转变能否保持住。这当然也不在我的掌控范围内，但我明白自己能掌控的是努力坚持自己做的事情。

第九部分

最近，我从一位记者那里了解到，我们与那些河道的清理没有丝毫关系。这是浙江省范围内的一个政府行为。我相信这花了很大一笔资金。最近，我注意到人们在这些河流的水域里用容器种植蔬菜等，河水看上去仍不是太好，但却比以前好了很多很多。

第十部分

说到水源，有个记者问我，说这些年来我们在山上清理垃圾，山上水库里的水干净了许多。坝体各边及水库周围原来布满了垃圾，但现在只偶尔出现零星垃圾需要捡拾。事实上，水库边上有个大家喜爱的烧烤点，人们常常留下成堆的垃圾。那地方总是令人扫兴。终有一天，我们会赢得这场战斗。总之，我想说的是那个记者试图把干净的库水归功于我，但我却没有接受。为什么呢？有

lot more rubbish than I ever did. They went to places around those waters that I have never been to. The supposed success that is attributed to me must always be given to all those hundreds of people who have made me look good.

Part 11

I started picking rubbish up with no big end-term plan. I just started as a remedy to my feelings of anger and helplessness about the state of the world environment. Very quickly it anchored down in my mind as something I would keep doing regardless of others' opinions or the weather and I would do it once a week and I would stick to the same area so that I could see (hopefully) some progress for my efforts. Gosh! I started using plastic bags and plastic gloves. Now that is a pretty environmentally unfriendly thing to do and I knew it but I also knew starting was more important than doing it wrong. Now, I am committed to keeping up this activity wherever I am in the world until I can no longer step outside my front door.

Part 12

April 10, 2014

Dear Friends and Family,

It has been quite a while since I have written about my project picking up rubbish on Da Luo Mountain here next to Chashan in the Wenzhou Higher Education District.

I didn't actually start back into my regular routine on Sunday mornings immediately when I arrived back here in the middle of

很多学生和其他人，他们捡拾的垃圾比我多得多啊。他们去过的那些水域，我从未涉足过。这所谓属于我的成功必须给予那些成百上千赋予我良好形象的人。

第十一部分

我开始捡拾垃圾时，并没有远大的终极目标。它始于修正我的愤怒情绪和对世界环境状态的无助情绪。不久之后，一个念头开始扎根我的脑海：那就是不管他人如何评价，也不管天气状况如何，我会一直坚持做下去，每周一次，我会坚守同一片区域，以便我能看到我努力的（可喜）结果。天哪！我开始使用塑料袋和塑料手套。那可是对环境不利的事啊，我明知这一点，但是我也知道动手去做比做错更重要。现在，我承诺，无论我在世界的哪个角落，我都会坚持这项行动，直到我年老体衰得无法迈出大门时为止。

第十二部分

2014 年 4 月 10 日

亲爱的朋友及家人：

关于在温州茶山高教园区附近的大罗山捡拾垃圾的项目，我已经撰写了好长一段时间了。

事实上当我在 2 月中旬回到温州时，我没有立即投入周日早上的日常工作中去，因为我脚痛。我花了好几个星期才恢复过来。

February because I had a sore foot. This took me a couple of weeks to get over. By then several people had asked me if and when I would be going back to work so I had half a dozen people with me even on that first day. That was enough to start the media machines and so there has been a constant stream of media people wanting to interview me ever since. At this stage there seems no sign that the stream is slowing down. Wenzhou TV News will be accompanying us this Sunday and it seems they intend to follow us for the whole morning which is quite a commitment for people like that. Up until now they usually want to get in and get out and on with the next job as quickly as possible. A couple of weeks ago there was a three-person team who had travelled from the capital city of Zhejiang Province for the day which meant they had to stay here overnight. They worked hard that morning lugging their gear up and around the mountain. I found it quite funny to often find a camera pointing in my face when I was on all fours clinging to a tree on the side of a hill while picking up rubbish. If you want a bit of a laugh then check out this link：http://www. wzu. edu. cn/Art/Art_6/Art_6_68709. aspx.

After this Sunday the next reporter who is booked in will come to my apartment on Wednesday afternoon. She doesn't sound quite so organised as she needed me to find my own interpreter but she is from China Tourism News which sounds a bit interesting. I guess the whole thing about these folk tracking me down is that there really is a lot of interest in any environmental story and this one at least so far is a bit of a success story.

那时已有好几个人在问我是否会回到大罗山捡拾垃圾的工作中去，什么时候开始，因此第一天开始工作的时候，就有五六个人跟我一起上山工作了。那已足够让媒体开始运转，自那之后，有一连串的媒体人员想采访我，在此阶段媒体似乎没有减少的迹象。温州电视新闻台本周将与我们一同上山，他们好像打算整个上午都跟着我们，他们这一举动的确是一种敬业精神的体现。到目前为止，他们通常是来去匆匆，马不停蹄地投入下一项工作。几个星期前，来自浙江省省会的一行三人为此前来采访，这意味着他们得在这里过夜。那天上午，他们拖着设备上山跟踪拍摄，工作相当卖力。在手脚并用攀附在山边的一棵树上捡拾垃圾时，我发现有摄像机对着我，这实在有趣。如果诸位想开心一笑，不妨查看以下链接：http://www.wzu.edu.cn/Art/Art_6/Art_6_68709.aspx。

这个星期天之后，我与另一位记者相约于星期三下午来我公寓里采访。她的英语组织能力似乎不够强，因为她需要我给自己找翻译，但有些逗乐的是，她竟来自《中国旅游报》。这些人之所以如此热衷于追踪我，我猜想是因为任何环保故事都会让人感兴趣，至少这个故事到目前为止多少算一个励志故事吧。

Many folk from Australia assume that I can arrange other squads of people to work in different areas but I am afraid that with language/communication difficulties and my general inability to deal with a huge cultural gap in behavioural and organisation expectations, that is simply not possible for me. However, it still happens! I have seen on one day recently that my old college still has a team working on the mountain from time to time. I saw another group that I have no idea where they come from or who might have organised them and I suspect that there are even more.

I do have plans to expand the project and today I took the first steps to make this happen. I spoke to a person from Oujiang College, (through an interpreter), and he is going to organise a banner and a few other bits and pieces for me. When I am all set which probably will not happen until next month I will go up to one of the approaches to the mountain and stop people on their way up and ask them to pick up a bag of rubbish while they are up there. If they agree to do it I will give them a cotton glove and a plastic bag to work with and ask them to wash the cotton glove and use it again one day. I do wonder if the plan will work but I have high expectations. I am quite sure there will be problems along the way of some sort or another. I will certainly need a translator alongside me to do this and maybe another helper as well. That will not be a problem so the real problem at this stage is that I don't see any problems. Ha ha! It reminds me of when I first started the work over 2 years ago. I had no idea what I was doing but bit by bit it all got sorted out and slowly we made significant progress.

很多澳大利亚人认为我可以在不同区域安排其他组别的人工作，但我却做不到，因为语言或交流存在困难，外加本人欠缺从行为、组织预期效果方面去应付文化鸿沟的能力。然而，奇迹还真就发生了！最近，我发现以前我工作过的学院仍有一队人马不时在山上捡拾垃圾。我还见到另一队人马，但却不知道他们来自何处，谁是他们的组织者。我觉得还有更多的参与者。

我确实有扩大该行动的计划，今天我为此迈出了第一步。我（通过口译）与瓯江学院的一个人交流过，请他负责为我制作一条横幅以及别的标识物。我也许要到下个月才能准备就绪，到时我将站在某个上山路口，让经过的人停下来，请他们在上山的途中捡一袋垃圾。如果他们同意这样做，我会给他们一双棉手套、一只塑料袋，让他们清洗手套以备将来再用。我不清楚该计划是否可行，但我期望值很高。我知道会碰到这样那样的问题。我当然需要一位翻译的帮助，也许还需要一个助手，这都不成问题。这个阶段的真正问题是我没有发现任何问题。哈哈，这让我想起两年前我第一次开始捡垃圾时的情景。我当时并不知道自己在做什么，但事情一点一点理顺了，我们慢慢取得了显著的进步。

It is a different environment on the mountain now but if you saw it, you would agree that there is a long way to go. A big part of the problem is that once people are in the habit of dropping litter it is not an easy habit to break when you are talking about many thousands of people. Many of these would come only once so they may not even be aware that on this mountain, it is no longer a good idea to leave their rubbish behind them. Perhaps you can imagine the situation where you have cleaned an area very well only to find that a week later it has been trashed so that it looks like a rubbish dump. I have lost count of the times this has happened and yet, there has proven to be a huge power in quietly bending to the task of cleaning it again. I am so lucky to have great companions along the way. Even though their faces change from time to time, they are still there.

Until next time, I wish you well with your magnificent obsession—whatever it may be.

要是你能看到，你会发现山上的环境变了模样，当然你会认为还有很长一段路要走。问题的关键是，人们一旦习惯于随地乱扔垃圾，你就是跟成千上万的人讲这个问题，那个习惯也不容易改掉。许多人只来一次，或许连在这山上乱扔垃圾是坏习惯的意识都没有。你或许能想象这种情形，你将某个区域清扫得干干净净，结果一个星期后又被弄得像个垃圾场。这种事儿到底发生了多少次，我已记不清了，然而，再一次埋头清理却蕴藏着一种巨大的力量，这已经得到了证实。我庆幸这一路有一群了不起的同行者。即使具体的参与者在时时变动，但那里始终有他们的身影。

下次再聊。不论诸位致力于任何高尚的事，我恭祝你们一切都好。

CHAPTER 5

An Altruistic Imperative

'The Altruism Coefficient' is a terminology that I think was originally heard in the movie 'The Box'. In this movie a stranger (who seems to be a person but is actually an alien of some sort), offers individual people a choice as a test of humanity's ability to be altruistic. The way I remember the movie was that either most people or everyone fails the test. In one of the scenes of the movie, the stranger explains himself to a USA National Security Agency (NSA) representative:

NSA Representative: 'You spoke earlier about the altruism coefficient.'

The Stranger: 'If human beings are unable or unwilling to sacrifice individual desires for the greater good of your species, you will have no chance for survival ... and my employees will be compelled to expedite your extinction. Clear?'

In another version of this saying which I might be making up myself or I have stolen from somebody else it goes:

'If human beings are unable or unwilling to sacrifice individual desires for the greater good of their species, the species cannot survive. The end is near.'

Stephen Hawking in his 'TED' talk has something similar to say that I believe is worthy of note. Here is the link if you want

第五章

利他主义之需

“利他系数”这个术语，我想我是在电影《杀人宝盒》中第一次听到的。在这部电影中，一个陌生人（看上去像人，但实际上是外星人）给每个人提供了一个选择，以测试人类的利他主义能力。据我回忆，在这部电影里大多数人或者所有人都通过这个测试。在该电影的一幕中，那个陌生人向一位美国国家安全局（NSA）的代表是这样解释的。

国家安全局代表：“你先前曾提到了利他系数。”

陌生人：“如果人类不能够或者不愿意为了自己种族的更大利益而牺牲个人的欲望，那你们将没有存活的机会……我的员工将迫不得已加速你们的灭绝。清楚了吗？”

这一说法的另一个版本很有可能是我自己杜撰的或者从别人那里抄袭过来的。它是这样的：

“要是人类不能够或者不愿意为了他们种族的更大利益而牺牲个人的欲望，那么该种族就不能存活。末日近在咫尺。”

斯蒂芬·霍金在其 TED 演讲中也有类似的说法，我认为这值

to go and listen yourself.

http://www.ted.com/talks/stephen_hawking_asks_big_questions_about_the_universe.html.

I have found it difficult to hear his exact words but my understanding of what he says is as follows:

He was talking about the future of the human race and he said: 'We are entering an increasingly dangerous period in our history. Our population and depletion of our resources of planet Earth are growing exponentially along with our technical ability to change our environment for good or ill. Our genetic code still carries selfish and aggressive instincts that were survival adapted in the past. Those same instincts will make it difficult enough to avoid disaster in the next 100 years let alone the next thousand or billion. Our only chance of survival is not to remain inward looking on planet Earth but to move out into space. Otherwise, our instincts will drive us to extinction.'

I am inclined to agree with Hawking. Even if we totally ignore arguments about climate change, even if we totally ignore the potential for disasters that seem poised to happen such as pandemics, collapse of the food production system, water shortage and consequent world war, —just the attitudes and expectations of government and people everywhere to have ever-increasing wealth, standards of living and population growth, logical thought must dictate that given a finite living space on our planet Earth, our species is doomed either to extinction or to periods of time with a lot of pain. When I say pain, I mean situations where millions and maybe billions of us die after long periods of protracted anguish

得关注。如果你感兴趣，如下便是该讲话的链接：http://www.ted.com/talks/stephen_hawking_asks_big_questions_about_the_universe.html。

我发现，要听懂他说的确切字眼很难，但对他所说的话，我的理解如下。

他在谈论人类的未来时说："我们正迈入人类历史上一个日渐危险的时期，随着我们改变环境的技术能力日益提高，这种能力非好非坏，我们的人口以及对地球资源的消耗正在呈指数方式增长。过去为适应生存而残留的那些自私和攻击的本性，我们的遗传密码仍然带着。正是这些本性将使我们在随后一百年里难以避免灾难，更不用说下个一千年或者十亿年了。我们生存的唯一机会不是只指望地球，而是迈入太空。否则，我们的本性将使我们走向灭绝。"

我非常赞同霍金的观点。即使我们完全忽略有关气候变化的那些论点，即使我们完全忽略发生潜在灾难的可能性，如全球性流行病、食物生产体系的崩溃、水资源匮乏，以及随之而来的世界大战等，单就世界各地的政府和民众对不断增加的财富、生活水平与人口增长的种种态度和期望，就目前地球上有限的生存空间来看，我们人类注定要么灭绝，要么历经数次痛苦的时期，才符合逻辑。我说痛苦，指的是以下这种情形：在我们过早地被死神带走之前，我们中数以百万计，或许数以十亿计的人要遭受长久的痛苦。我

before death takes us prematurely. When I say prematurely, I mean that most of us will die well before we reach even 50 years of age and many more will die as children.

There seems to be a great need for altruism to ensure the survival of our species. To put this plainly, maybe we need to accept a lower standard of living than what many of us expect. Maybe we need to eat a lot less and in particular many of us need to eat a lot less meat. Maybe we need to plan for the GDP of our countries to shrink a lot. Maybe we need to stop population expansion beyond what we have now and even consider reducing population in time. Maybe we need to learn to put up with the car/computer/mobile phone/private jet that we use for an extra year or two before we buy the latest version. Maybe we should pay full price for any non-replaceable resources we use from our planet. Maybe we should do all these things sooner rather than later. Nobody will blatantly say that they will not do these things if in the same breath they are forced to admit that if they don't do these things, the next generation will have to sacrifice a lot more. Nobody will openly admit to being so selfish. In the meantime we prefer not to think about it at all.

What do I mean when I say we should pay the full price for things we use? I mean that when we buy our chicken meat or any other meat that is produced on factory farms we should pay the full price for cleaning up the polluted environment around these farms rather than the artificially deflated prices that we have got used to. Maybe we should also admit with grace, that we are happy about forcing creatures to live and die in these cruel

说“过早”，指的是我们当中的大多数人远不到五十岁就会死去，还有更多的人会夭折。

要确保人类的生存，我们也许急需那种利他主义精神。简单说来，也许我们应该接受比我们期望值更低的生活水平；也许我们应该吃得更少，特别是少吃肉类；也许我们应该制订大力减缩各国国民生产总值的计划；也许我们现在应该阻止人口再增长，甚至考虑及时减少人口；也许我们得学会忍受已有的汽车、电脑、手机、私人飞机一两年，再购买最新版；也许我们应该为任何不可替代的资源支付全价。也许我们应该及早实施这些措施，因为不实施这些措施，下一代将会付出更大的代价。若人们被迫承认以上这一点，没有人会公然宣称他不愿意采取这些措施。没人会公开承认自己自私。因此，我们宁愿掩耳盗铃。

我说我们应该全额支付自己使用的东西，这是什么意思呢？我的意思是，在我们购买由饲养场生产的鸡肉或其他肉类时，我们应该为清理这些饲养场附近被污染的环境的所有费用买单，而不是支付我们习以为常的市场价格。也许我们还得大方承认，为满足我们对美食的欲望，我们乐意看到这些动物在那些残忍的饲养

farms for the sake of our gastronomic desires. Maybe we should admit that this attitude is totally at odds with our feelings for our pet dog, or pandas, or koala bears. Maybe we should admit these attitudes are totally illogical for a human being with the ability to think.

What else might I mean by this idea of paying the full price for things we use? One could write a book about things that don't cost nearly as much as they should if we factor in the cost of cleaning up the pollution caused by moving food incredible distances, or the cost of cleaning up and disposing of packaging that is needed for the sake of our convenience, health and safety.

Many people act towards me as if I am some sort of holy altruistic saint. They are wrong. They just haven't figured out yet that I am as selfish and self-involved as many other people but maybe, just maybe, I take a longer-term view of many things. Maybe there are too many people looking at things and planning for things with a three-year plan. I could give as one prime example the government of Australia and through them, most people of Australia. I look at the government of China and at least they often seem to be working with a ten-year plan in mind. I am pretty ignorant to the intricacies of both of these governments so pull my arguments to pieces if that is your desire. You can even seek to change my opinion if that is important to you.

I am by nature a very selfish man. I have not even been a very good father or husband because I am so selfish. Maslow's hierarchy of needs puts survival as the most basic and that is the position I am operating from in taking the actions that this book

场里生存与死亡。也许我们还得承认，这种态度与我们对宠物狗、熊猫以及考拉的情感是完全对立的。也许我们还得承认，这种种态度对能思考的人类来说是完全不合逻辑的。

我说我们应该全额支付自己使用的东西，我还有其他别的意思吗？有关清理长途运送食品所造成的污染费用，以及清理和处置我们为方便、健康和安全而产生的包装垃圾，你能写出一本有关物品与实际价格不成比例的书。

不少人将我看成一个利他主义的圣徒。他们错了。他们只是没有弄清楚我跟其他人一样自私，一样以自我为中心，但是也许，仅仅是也许，我看待许多事物的眼光要更长远一点而已。看待事情并制订出其三年计划的人也许并不少。我可以拿澳大利亚政府作为典型例子，通过该例可见多数澳大利亚人的思维。在我看来，中国政府往往在工作中至少考虑到的是十年规划。我对两国政府运作的复杂细节一无所知，如果你愿意，不妨把我驳个体无完肤。你如果非常在意这一观点，不妨试着改变我的看法。

就本性而言，我非常自私。我因如此自私而不能成为一位称职的父亲或丈夫。马斯洛的需求层次理论将生存置于最底层，那正是本册子所涉及的我所采取行动的出发点。在这个维系我生存

is about. I want to survive and live for as long as I can on a planet that supports my living. I could be dead before this book has ever seen the light of day but before then, I want to run along tracks that are not littered with rubbish. I want to breathe air that is clean. I would really like to be able to stop running when I am thirsty and take a drink from a stream without using a filtering device, but there is almost nowhere in the world where I can do that now. Don't you find that appalling? I can remember being able to stop at any river that I came across and so long as there were not cows or sheep grazing up stream, I could take a drink. Most children in our world today will never ever have that memory and most of them will never have the experience—ever.

If just that one idea does not turn you into a person wanting to do everything you can possibly do to change the current situation, then maybe you do need some altruism transference into your character. Maybe you do need to become selfish enough to start thinking of the longer-term consequences of your action or inaction. Hey! Maybe I am totally wrong and maybe it is right to just not care!

What do we do when we come across the ultimate attitude of non-caring? Sometimes one comes across people who really don't care what happens or what damage they may cause anywhere. They generally don't think much but perhaps they figure that they will soon be dead and so why should they care? Often their superficial argument is that the person next to them doesn't care, then why should they?

的星球上，我想尽可能长久地生活。在本册子面世之前，我也许已经死了，但是在我去世之前，我想在那些没有垃圾的小道上跑步。我想呼吸新鲜的空气。我口渴时，能够停下来，喝上一口不需要过滤的溪水，可在当今世界的几乎任何地方，我都难以如愿。难道你不觉这很可怕吗？我仍然记得自己从前驻足没有牛羊喝水的河边的情形。当今世上的大多数孩子不会有这样的记忆，也不会有这样的经历——永远不会。

要是这一想法还不能让你变成一个竭尽所能改变现状的人，那么你的性格中确实需要植入一些利他精神。也许你确实需要变得足够自私自利，该考虑因自己作为或不作为而引发的长期后果。唉，也许是我完全错了，也许漠不关心才是正道。

当我们碰到那种漠不关心的强硬态度时，我们该怎么办呢？有时候，我们确实会碰到这类人，他们全然不顾会发生什么或者其行为会给什么地方造成了破坏。他们一般不太思考问题，也许他们认为自己不久就会死去，那么又何必在乎呢？他们通常肤浅地认为周围的人都不在乎，那他们干吗要在乎呢？

I can understand this type of thinking. If one has no connection to the larger world outside of self-interest this is perfectly logical. If one has absolutely no spiritual understanding or belief system then this attitude makes good sense. If one has never felt closeness with the earth or felt the power in the forces of nature as being a part of themselves, then indeed, why would they be bothered with anything that did not offer immediate benefit to themselves?

I don't want to delve into spiritual matters in this book and I do not have the skill to describe the intimacy, the power, the largeness of knowing one is not just a small part of our Earth. I know, even if I can't explain it very well, that each one of us is much more than that.

It is my contention that the best way to meet with the attitude of non-caring is to keep caring. I think it fits in nicely with the philosophy of non-violence and it works. It doesn't often work immediately so one has to be able to take a long-term view of things and have a ton of patience. Maybe if you have great powers of persuasion you could convince the person who really doesn't care about the longer-term good of their environment that it is in their own self-interest to change their attitude. If that power is in your skill set then I beg you—GET TO WORK. In the meantime, I will plod along quietly trying to give and care as much as a selfish man with a long-term attitude can do.

我能理解这种想法。如果一个人除个人利益外，与外面广阔的世界毫无联系，那么这种想法是相当合理的。如果一个人没有丝毫的灵性理解或者信仰体系，那么持有这种态度也是说得通的。如果一个人与地球从未有过亲近感，或感受过自然之力，那么何必要为那些不曾给他带来直接利益的事情苦恼呢？

在本册子里，我不打算深究精神层面的东西；我明白人绝非地球上小小的一部分，却无力描述出其应有的亲切感受、活力以及宽厚。我明白，即使我描述得不够到位，但我们每个人远不只是地球的一小部分。

我坚持认为，克服漠不关心态度的最有效办法是不断地关注。我认为这与非暴力哲学完全吻合，并且行之有效。它通常不会立竿见影，因此我们需要从长远的观点去看待并保持足够的耐心。如果你有雄辩的口才，你也许能够说服那些对周围环境的可持续发展漠不关心的人，让他们意识到改变其态度也关乎其切身利益。如果你有那种能力，我恳请你——**行动起来吧**。与此同时，我将作为一个具有长远眼光的自私者，默默前行，尽力奉献与关注。

CHAPTER 6

You Are Not Alone

I once had a phone call from a guy complaining that he had been picking up rubbish but he had stopped doing it for a variety of reasons. It is as a result of that phone call that I am including this chapter because I thought it was sad that he had stopped doing what was a worthwhile thing to do. Since that time I have met many people with a not dissimilar story to tell.

I might have stopped picking up rubbish at one point, but I happened to do a bit of internet searching and I found out that I was not alone in the world. Other people all over the world were doing the same thing. Some people do amazingly difficult work trying to clean up our environment. Other people do amazingly difficult work trying to stop our environment being polluted. Millions of people just like me (and maybe you) are doing little bits to help our world be the beautiful place it should be, and it is worth remembering them so that you do not feel discouraged and give up.

First of all, other people voluntarily pick up rubbish. They may not have matching caps or fancy uniforms and they may be working alone, but they exist all over the world, and they also exist in China. I have seen them working and I have seen the results of their work. In Australia on some of the highways you can see

第六章

你并不孤单

我曾接到过一个电话，来电者抱怨说他之前一直在捡拾垃圾，但由于各种原因现已放弃。正是由于那个电话，我才把这一章写了进来，因为在我看来，他不再做这件值得一做的事情实在是一大憾事。自那时起，我见到过很多有类似经历的人。

我曾一度可能也不再捡拾垃圾，但我碰巧在网上搜寻，结果发现在这个世界上不止我一个人在行动。世界上还有其他人在做同样的事情。有些人为了清理我们的环境做了相当艰苦的努力，有些人为了阻止我们的环境被污染也做了相当艰苦的努力。为恢复美丽世界的本来面目，数以百万计像我（也许像你）一样的人都在尽其绵薄之力。他们值得你铭记，让你不再沮丧，不再放弃自己所做的事情。

首先，其他人在自觉地捡拾垃圾。他们没配备清一色的帽子或者花哨的制服，他们也许是独自工作，但包括中国在内的世界各地均能见到他们的身影。我见到过他们的工作身影，也见到过他

signs where local community organisations proudly take responsibility for sections of the highway to keep it tidy and free of litter. They do this work voluntarily in many cases. In other places the local government has seen fit to pay somebody to do it. It is always worth remembering that their pay comes from the taxes that everybody pays. If there was no need for them to do this job then the taxes could be spent on something to make our lives even better.

Then there are the people who are always thinking about their impact on the environment by how they do their food shopping. Many of these people almost exclusively buy only local produce as it seasonally becomes available. I've never met anybody in Australia or the UK who do this because they get so used to supermarkets with a huge variety of products available to satiate our senses on a daily basis. There are many ways in which this costs our environment dearly but we almost never consider them. I have heard of people who try to live without creating any rubbish at all. They recycle everything they use so that their rubbish bin is never put out to be transported to land fill. This is an extreme way to live and for many of us, we cannot even think how this would be possible and yet they do it. How? I don't know but the idea has made me think. One day I hope to meet some of these people.

Another way in which people think about the environment when they are shopping is in how many plastic bags they are using and if they are using these bags, how often do they get used before they finally reach the trash. Plastic bags are one of the big causes of

们的劳动成果。在澳大利亚的高速路上,你可以看到一些标识牌,上面写着当地社区的某一组织自豪地负责保持那些路段的清洁。在很多情况下,他们自愿从事这些工作。在其他地方,当地政府可能雇人做这项工作。请记住,清洁工的薪酬来自民众的纳税。要是不雇人清理道路,那么这笔税款就可以用于改善我们生活中的其他事情。

其次,有些人总在食品购买方式上下功夫,来减少自己对环境的影响。这类人几乎只买当地的时令农产品。我在澳大利亚或者英国从来没见过这样的人,因为那里的人们习惯在购买超市那些琳琅满目的产品,以满足其每天的口味。我们如果留意,就会发现存在着很多让我们的环境付出高昂代价的生活方式,但我们大都不在意。我还听说有人尝试以不产生任何垃圾的方式生活,他们循环使用所有东西,他们的垃圾也从来不需要送往垃圾填埋场。这是一种极端的生活方式,对我们多数人来说,简直连想都不敢想,可他们就是这样做了。怎么做到的呢?我不知道,不过这个主意倒是引起了我的思考。我希望有一天可以见识一些以这种方式生活的人。

人们思考环境的另一种方式是:当他们购物的时候,他们要用多少个塑料袋;如果要用塑料袋,这些塑料袋在变成垃圾之前能用

both unsightly pollution and chemical pollution of our earth and our water. We seem to be addicted to their convenience and how cheap they seem to be. That addiction crosses boundaries into almost every country in the world. I think the only way we can constrain the addiction is to make the real cost of cleaning up the land we grow our food in and cleaning up the oceans we take our fish from, reflected in the upfront cost of using these plastic bags.

Other people think about creating plastic bags and packaging that decompose quickly and without damaging our environment. I know that any progress in this area must be a good thing because I cannot ever see us totally weaning ourselves from our addiction to the convenience of packaging and disposable products.

As I write this chapter I am currently in the UK. Yesterday I was out running and saw an awful mess of rubbish around an empty bin. The bin was obviously emptied regularly by paid workers, but for some reason this other rubbish had accumulated in the area around the bin and the workers did not see it as their responsibility to pick it up. I spent about fifteen minutes cleaning up the area. Nobody helped me and nobody saw me do it, but I was not alone. I know that at that very moment, somewhere else in our world, somebody was doing the same thing or something very similar. We all make a difference and it all adds up eventually. If you are doing it just for the recognition though, you may be disappointed. I got satisfaction from my work just by looking back for a few seconds and thinking to myself that was a job well done. That is all. That and the fact that I know I am not alone.

You too are not alone. It may be a demeaning thing to pick

多少次。塑料袋是污染我们的土地和水源的重要来源之一，既造成不雅观，又导致化学性污染。我们似乎离不开它们带来的便利及其廉价的成本。这种依赖跨越国界，进入世界各个角落。我想，抑制这种依赖的唯一可行方式，就是一并计算清理我们种植食物的土地、清理我们捕鱼的海域所花的实际成本，让实际成本体现在使用这些塑料袋的前期成本中。

也有些人在考虑生产出可快速分解、对环境无害的塑料袋和包装材料。这一领域取得的任何进步都是一件好事，因为我知道我们未必能完全放弃对包装产品和一次性用品的依赖。

撰写本章内容时，我正在英国。昨天出去跑步，我看到一个空垃圾桶周围有令人恶心的垃圾。这垃圾桶显然由工人定期清空的，但出于某种原因，周围的垃圾不断堆积，工人们却没有将收拾那些垃圾纳入自己的职责范围。我花了十五分钟清理完那片区域。没有人前来帮忙，也没有人看见我做这个事情，但我并不是一个人在行动。我知道就在那时，地球上的其他地方，有人在从事相同或者相似的事情。我们大家都做一点改变，加在一起，最终汇成巨变。要是只为得到人们的认可才去做，你就会失望。我回头凝视几秒钟，心想自己干了一件值得干的事情，就心满意足了。仅此而已。我知道自己并非一个人在行动。

up other people's rubbish when seen through some people's eyes. Through other eyes it is the act of somebody who really cares about their environment. It is the act of somebody who cares about quality of life. It is the act of somebody who thinks about the future. It is the act of somebody who can think about the long-term impacts of both their actions and the consequences of lack of action. It is the act of somebody who is willing to take care of the little things because they know that little things add up to become big things. You can be that person.

People may laugh at you behind their hands or behind your back. They may laugh in your face and even tease you. It may be raining or snowing or hot or cold and uncomfortable. You may be tired or hungry or lonely. Your legs or your back may be hurting. You may be thinking you would rather be somewhere else. You may be in a state of emotional crisis because this happens to everybody from time to time for different reasons. None of these things matter at this point quite as much as they otherwise would. Why don't they matter so much? Because what you are doing is worth doing and you are not alone.

In my classrooms while teaching English in China I am careful to never proselytise my point of view on any subject. At the same time I am aware that it would be impossible for students not to pick up on the fact that I have some degree of passion about the world we live in. At one stage in this teaching career of mine I collected some quotes that students wrote. These particular students were not good at English and I would not have rated them as having very good environmental awareness but I was wrong.

你并非一个人在行动。在他人来看，去捡别人丢下的垃圾也许是一件不体面的事情。但通过另一些人的眼光来看，这正是那些关心周围环境的人应有的行为；这正是关心生活质量的人应有的行为；这正是关心未来的人应有的行为；这正是那些能够长远地看待自身作为或者不作为后果的人应有的行为；这正是那些愿意关注小事的人应有的行为，因为他们明白积少成多、集腋成裘的道理。你也可以成为那样的人。

人们也许会掩面嘲笑你，会在背后嘲笑你，或者当面嘲笑你、作弄你。也许正下着雨，下着雪，或天气炎热，或寒冷刺骨，令人难受。你也许会疲惫、饥饿或者孤独无助。你的双腿或者脊背也许正疼痛着。也许你会希望自己宁愿在别的什么地方。也许你正处于情感危机中，因为这种情况会出现在每个人身上，只是原因各有不同而已。这一切在别的时候兴许重要，但在此时却无关紧要。为什么无关紧要呢？因为你在做的事是值得做的，况且你并非单打独斗。

在中国讲授英语时，我总是小心翼翼，绝不在任何话题里宣扬自己的观点。与此同时，我意识到学生一定能感受到我对这个世界的那份热情。在教学生涯的某个阶段，我收集了一些学生写的话语。这群学生的英语不好，我认为他们的环保意识也不强，但是

The following selection of quotes from some of these students proves how wrong I was. My point in including them here is that you and I do not know what other people are thinking. A lot of people are probably thinking similar things to you but maybe they need you to get up and lead the way in doing whatever you can do.

Humans are the owners of the earth, but also the destroyers of it and the final victim.

Kevin. 陈斌武

Please protect the environment. Don't let our tears become the last drop of water.

Arthur. 周伟民

Only by changing the way we treat the environment can we get along well with it. Only by saving the environment can we save ourselves.

Warren Pan. 潘万鹏

To improve the environment means to improve our lives. We should plant more trees and flowers around us. We shouldn't cut them down. We should stop factories from pouring waste water into the rivers and waste gas into the air. Whenever we see litter on the ground, we should pick it up and throw it into dustbins. It's our duty to protect the environment.

Lincoln. 林浩洁

我错了。下面引用的一部分学生的话语也证明了我错得一塌糊涂。将他们的话语摘录于此的用意是：你我并不了解别人的想法。也许很多人都在思考差不多一样的事情，也许只需要你起身带头，一起去做你能做的事而已。

人是地球的主人，也是地球的毁灭者，也是最终的受害者。

——凯文（陈斌武）

请保护环境。别让我们的眼泪成为地球上的最后一滴水。

——亚瑟（周伟民）

只有改变对待环境的方式，我们才能与其和谐相处。只有拯救环境，我们才能拯救自己。

——潘华伦（潘万鹏）

改善环境意味着改善我们的生活。我们应该在自己的周围多种植花草树木。我们不应该滥砍滥伐。我们应该阻止工厂将废水排入江河中，将废气排入大气里。不论何时看到地上的垃圾，我们都应该捡起来放进垃圾桶里。保护环境是我们的责任。

——林肯（林浩洁）

I hope scientists can find ways to solve the serious problems. Maybe then we could look forward to seeing the clear sky, clean rivers, and beautiful parks again.

Braden. 赵斌生

If we plant a lot of trees, if we clean up the rivers, if we eat more vegetables and less meat, if we walk instead of driving—the environment will be more beautiful.

William. 傅威浩

My suggestion is we should ride bikes more often instead of driving cars. By riding a bike, we can not only exercise our body but also protect our environment.

James Huang. 黄东亮

If the population continues to rise at present rates, in a few years there won't be enough food.

Liam. 项灵敏

It is shameful to gain a benefit while damaging the environment.

Kayle. 李翔

If each of us does something, I believe in the future environmental problems will be improved greatly.

Layla. 何飞娇

我希望科学家们能找到解决这些严重问题的方案。如果是这样的话，我们或许可以期待再度见到蔚蓝的天空、干净的河水以及美丽的公园。

——布雷登（赵斌生）

如果我们大量植树，如果我们清理河道，如果我们多吃蔬菜少吃肉，如果我们安步当车，环境就会变得更加优美。

——威廉（傅威浩）

我的建议是，我们应该多骑车而不是多开车。骑车，我们不仅可以锻炼身体，而且还可以保护环境。

——詹姆士·黄（黄东亮）

如果人口以目前的速度继续增长，几年过后，食物将会匮乏。

——利亚姆（项灵敏）

以破坏环境而获利的行为是可耻的。

——凯勒（李翔）

如果我们人人都尽力而为，我相信将来的环境问题将会大大改善。

——雷拉（何飞娇）

As humans, we are all living under the same clouds and walking on the same earth. There is only one earth and there isn't much of it. So let us work together to protect the environment.

Bryant. 金斌

我们人类都生活在同一片蓝天下，行走在同一个地球上。地球只有一个，而且不堪重负。让我们行动起来保护环境。

——布莱恩特（金斌）

CHAPTER 7

Power and Progress

Where does power come from? How does one find one's own 'glorious obsession'? I am quite sure that I don't know. I have found however, that even I have personal power and through my obsession with rubbish littering beautiful places, I have been able to use it. Why or how have these seed ideas grown into what has happened here in Wenzhou on Da Luo Mountain? There have been a few clues I have picked up along the way for me and here they are.

The first would be character. Just as in teaching, one cannot become a good teacher by simply copying the best teachers. We can learn from them but we must become aware of and use our own character and personality. Even with all our flaws and inconsistencies, everybody has some strengths and it is on these strengths that the best of our actions will lie or be aided by. One of the keys to my success here has been that I have been obstinately persistent and this is a major part of my character. Whether it has been ultra-marathon running, success as a clarinet player, or selection into the Australian Army Special Forces, I have achieved these things through the ability just to keep going even when the task ahead seemed impossible. I can only describe this trait as obstinate persistence. It can be good or bad but in this

第七章

力量与进步

力量来自何处？人们如何找到属于自己的那份“光荣的痴迷”？我确实不知道。然而，我发现连我也拥有这种个人的力量——痴迷于捡拾污染美丽地方的垃圾，我一直有效地运用这种力量。是什么原因或者什么方式让这些思想的种子开花结果，从而成为温州大罗山上所发生的一切？一路上我自己“捡拾”了几个线索，现奉如下：

首要原因应该是性格。恰如教学一样，一个人如果一味效仿优秀教师，是成不了好老师的。我们可以向他们学习，但我们必须意识到我们自己的性格和个性。即使我们有种种不足和欠缺，但人人都有些优点，我们的最佳行动得以依靠或辅佐的正是这些优点。我在温州取得成功的秘诀之一是我一直坚持不懈到了倔强的地步，这也是我性格的主要组成部分。不管是参加超级马拉松赛，成为一名成功的单簧管演奏者，还是入选澳大利亚特种部队，我获得这一切凭的就是不断坚持，哪怕前面的任务看似不可能完成。我只能把这种品质描绘为一条路走到黑。这种品质的优劣均存，

situation I am obviously focused on the good it can accomplish. The bad aspects of this character trait can be pretty ugly so I'm going to ignore them for the purposes of this book.

The second reason for this small success with improving the environment here in Wenzhou would be that I had the right idea at the right time in the right place. I place this reason second because I believe that it is secondary to persistence. There can be no denying that being a foreigner has meant that anything unusual I have done here has been noticed a lot more than if I had looked like a Chinese person. I have always been very aware of this phenomenon and realised that I could use it to advantage.

Another part of being in the right place at the right time is that I feel that Wenzhou people are thinking people generally. With their famous business acumen there comes an ability to think and to be more aware of a situation and what is developing. This ability and the effect it has had on my efforts should not be underestimated. Combined with this is a general world awareness that we must take better care of the land we live on if we want to live well.

The third reason I have chosen is the force and power of habit. Most people know about bad habits but beyond that they have never thought about habits and they have never studied how to use habits to their best advantage. For example a bad habit may be hard to break but it is relatively easy to start a new habit. If you start enough new habits then there is no room in your life for the bad habits so they quietly and easily disappear.

但在此情景中，我显然集中关注的是它能成事的优点。这种性格的劣处可以说糟糕到极点，但因本册子本身的意图，我将略去不谈。

我能成功改善温州环境的第二个原因是天时、地利、人和。我把这个原因放在第二位是因为它应该从属于毅力。不可否认的是，作为一个外国人，我在这里做的任何不寻常的事情所受到的关注程度会远高于中国人。我一直留意这种现象，并意识到可加以充分利用。

天时、地利、人和的另一方面，总的说来，我觉得是温州人善于思考。其敏锐的商业头脑造就了他们对局势的洞察能力与对形势的把握能力。这些能力及其对我努力的影响是不可低估的。与此相连的是对世界的一个总体意识，即我们想要生活得更好，就必须更加关爱我们生活的这片土地。

我选择的第三个原因是习惯的力量。大多数人对坏习惯有所了解，却不曾对习惯有更多的思考，也从没有研究过如何最有效地发挥习惯的作用。比如，坏习惯很难纠正，而养成一种新的习惯则相对简单一些。如果你养成了足够多的新习惯，你的生活就不会给坏习惯留下余地，而坏习惯就容易悄无声息地消失。养成良好的习惯通常是一种有意识的决定；必须自觉意识到好习惯对我们的重要性，我们才能养成和保持这种习惯。练习留心自己的情感、思想

Starting a good habit will usually be a conscious decision and it must be developed and then maintained by a conscious awareness of why it is important to you. Practising being consciously aware of our feelings, thoughts, or any situation, can lead to personal insights that are always good for us. After developing a habit that we have consciously and purposely chosen, it has a power that seems to come from within the habit itself. We only have to contribute a little energy to maintain it. This means we have more energy for other things which is a pretty nice thing to have.

While I have been here living in the Wenzhou Higher Education Zone my habit has been to go picking up rubbish every Sunday morning and I leave a set point at 8 am. People still ask me whether I will go if it is raining. I personally don't think I can maintain a habit like this if I allow the weather to dictate if I will do it or not. I don't even think I can maintain a habit like this if I allow my personal feelings to dictate if I will do it or not. If a habit is to become a power force in our lives there can be almost nothing that interrupts it. In what is now more than two years I joke with my students that there have only been two times that I did not maintain this habit. On one occasion I was so sick and weak from vomiting all night I was simply unable to make the trek up the mountain, but I still took all the equipment to the start point at the regular time. On another occasion I wanted very much to attend a table tennis tournament so I arranged that a student would be the leader for that day. I really should not have allowed that exception to occur but it happened. It will probably happen again because there are other passions in my life

或者任何处境，可以练就对我们有百利而无一害的个人洞察力。当我们自觉而有意选择的习惯得以养成，那它就似乎拥有一种来自内在的力量。我们只消稍作努力便可保持。这也意味着我们为做其他更好的事情腾出了更多的精力。

在温州高教园区，我的习惯是每个星期天去捡拾垃圾，我固定在八点钟出发。人们仍然会问我下雨天是否还去。我认为，如果让天气决定我去还是不去，那我就不可能保持这样的习惯；我甚至认为，如果让我的个人情感来决定我去还是不去，我也不能保持这样的习惯。如果习惯能成为我们生活中的强大力量，那任何东西都无法打断它。在两年多的时间里，我跟学生开玩笑说，我只有两次没有保持住这个习惯。一次是我病得很重，整晚呕吐导致我虚弱得无力上山去，但我还是在规定的时间将所有设备带到了出发点。还有一次是我很想去看一场乒乓球赛，于是就安排一个学生作为那天的队长。我真不应该让例外现象发生，但它偏偏就发生了。这种例外现象也许还会发生，因为除捡拾垃圾外，我在生活中还有别的爱好，但我明白不可以轻易就让这些例外现象出现。为什么？因为我不想丢失习惯的力量。实际上，还有其他一些例外现象。当我离开温州半年后回来的时候，我因脚痛而推迟到两个

apart from picking up rubbish but I know that these exceptions must not happen easily. Why? Because I don't want to lose the power of the habit. In reality there have also been other exceptions. When I first arrived back in Wenzhou after being away for 6 months I had a sore foot that stopped me starting the work for two weeks. On other occasions government holiday arrangements have meant that I had to work on some Sundays. On these occasions I always went on an alternate day. The point is that the habit became so deeply ingrained that it was relatively easy to go even though it was cold, windy and snowing. It became relatively easy to go even though I was tired and my back hurt. It became relatively easy to go even though I had only got back to Wenzhou from doing a Chinese exam in Shanghai a couple of hours previously. It still took effort to go in these situations but it didn't take as much effort as it would have if I didn't have the habit firmly engrained in my life.

Even now I wonder if I still underestimate the power of using habits to achieve what I really want.

The fourth factor I will now consider is passion. Am I passionate about the environment? I guess I am but am I more passionate than you or anybody else? I doubt it. When you examine most people and what they think about environmental concerns and you really get down to the basics of air, water, and food, everybody starts to get passionate. They start to get really passionate very quickly when you take away their air, EXCEPT: If you take it away slowly and introduce pollutants slowly then you can trick them into thinking that it is normal and an acceptable price to pay for

星期后才开始工作。有些时候，法定假日的安排意味着我得在星期天上班。遇上这种情况，我总是改到其他时间上山。关键在于习惯一旦深深扎根，即使寒冷、刮风、下雪，即使我疲惫不堪、腰酸背痛，即使我几个小时前才从上海参加汉语考试回到温州，去劳动的事也会变得相对容易。在这种情形下去劳动仍是件费力的事，但却没有想象的那样费力，因为我让这种习惯植根于我的生活之中。

即使现在，我还不知自己是不是低估了习惯在我获取真正所需之物时发挥的力量。

我接下来要考虑的第四个原因是激情。我对环境保护有激情吗？我想我有的，但我比你或者别人更有激情吗？我表示怀疑。若你审视大多数人以及他们对环境问题的种种关注，当你真正触及诸如空气、水源、食物等基本要素，任何人都会变得激情满怀。如果你将他们的空气夺走，那他们就会立马义愤填膺起来，**但是**如果你慢慢将其拿走，慢慢引入污染物质，温水煮青蛙，你可能会让他们误觉这些都再正常不过，因为要获得就应该要付出。人们认为那是没关系的。如果我们慢慢引入这一新的观念、新的训练模

having a job. People think it is then OK. We can get used to almost anything if we introduce the new idea, the new training, the stifling pollutants in the air, the poisons in the food, or the muck in our water, slowly enough. This is what has happened in our world in too many places and I think that this situation needs to be noticed. Then we have to start paying the price to fix the situation. I suspect that I have now shown my true colours. I am passionate about our environment and if you study success manuals you will find that passion is a key to success in most things.

The fifth factor is discipline. I spent a total of 23 years in the Australian Army. In spite of the imposed discipline factor which often correlates well with self-discipline, it does not necessarily mean that I have great self-discipline and my own assessment of my qualities in this area would rate my self-discipline as sadly lacking most of the time. However, I do on occasions demonstrate that I can be disciplined at least some of the time. It takes some discipline to develop a habit. It takes some self-discipline to keep at a task when the going gets tough. It takes some self-discipline to remind oneself that what other people think is not my business. What I do, IS my business. Self-discipline has been described somewhere as the pathway to success. I remember reading in another place that self-discipline gives you the freedom to succeed. I could argue that it gives you freedom—period. Freedom is a fallacy without personal responsibility and discipline. That could be another book topic if anybody is interested.

The sixth factor that I believe has given me the power to go

式、空气中令人窒息的污染物质、食物中的有害成分、水中的污物，差不多任何东西我们都会习以为常。在我们这个世界上，发生这类事的地方实在太多，我认为这种情形应该引起注意。我们不得不花钱来扭转这种局势。我想我已经表露了我真正的意图。我对环境充满激情，如果你查看一些成功指南，就会发现多数事情之所以取得成功，激情是个关键。

第五个因素就是纪律。我在澳大利亚部队共计待了二十三年。尽管有强制性的纪律因素，这通常与自律相关，但这未必意味着我拥有极强的自制力，我对自己在这方面的评价是，多数时候缺乏得可怜。然而，有时候我确实能够约束自己。要养成某种习惯需要一些自律。当某任务变得艰难的时候，这就需要自律去坚持。提醒自己他人的想法跟我无关是需要一些自律的。我做什么是我自己的事情。自律一直被描述为走向成功之路。我记得在某处读过一篇文章说，自律赋予你成功的自由。我可以证明它给人自由——没错。没有个人责任和纪律，自由就是一种谬论。如果有人感兴趣，这也可以成为另一本书的话题。

让我坚持捡拾垃圾并取得一些成效的第六个因素一直是友谊。

on and achieve something simply by picking up rubbish, has been the unintended reward of friendship. When people know they can depend on you they are much more willing to be friends and that friendship comes with benefits that no money can every buy. Genuine smiles, handshakes and compliments don't come cheaply but the value they contribute to my life are worth more than anything I might have paid to receive them.

Every week I would go and follow the same route at the same time. Sometime I was on my own and sometimes I would have just one person with me and often the same person would appear for a long time. There are different names that come to my mind when I think of them but they were all precious to me and they were all great companions. Here are some of the English names of people whom unfortunately I do not know their Chinese names: Catherine, Bruce, Euken, and Tom. Then there is Clearmay (刘明媚), Leah(裴莉亚), Michelle(谢朱琳), Alice(郑赛男), and the glorious Betty(张玲燕).

In addition to the great benefits that friendship brings there was and is something really funny that I have felt happens as a consequence of relentlessly sticking to the task. It was as if I was empowering myself to be able to do many other things. The act of maintaining that habit or that discipline was in itself a thing that gave back power. It made me realise that there was almost nothing I could not do if I applied myself with this energy. This was and is a reward that is worth all the effort of getting out once a week and picking up rubbish. Any recognition I have received is a small reward by comparison. Whenever I have found

这是一个意外回报。当人们认识到他们可以依靠你,他们更愿意跟你做朋友,那种友谊带来的好处是用金钱买不到的。真诚的微笑、热情的握手和衷心的赞美并非唾手可得,它们赋予我生活的价值比我花钱买到的任何东西都大。

每星期的同一时候我都会走同一条线路。我有时是独自一人,有时仅有一个同伴,这人通常会坚持很长一段时间。每当想到他们,我大脑中会出现不同的名字,这些人我都非常珍惜,他们都是我的挚友。下面便是其中一些,不幸的是一部分人的中文名字我没能知晓:凯瑟琳、布鲁斯、于肯、汤姆、柯丽梅(刘明媚)、李(裴莉亚)、米歇尔(谢朱琳)、爱丽丝(郑赛男),还有了不起的贝蒂(张玲燕)。

除了友谊带来的好处外,由于我坚持去做这件事,其间我发现一件非常有趣的事。我仿佛充满能量,可以做很多其他事情,可见保持那种习惯或纪律的行为本身就能回馈力量。这使我意识到,要是我利用这份能量,几乎没有我做不成的事情。这是一份回报,

myself not acting as I would really want to, the memory of this discipline comes back to me and I know, I can do and act in a way that I really want to. I need no longer be governed by transient feelings that might lead me astray from my intended course of action. This is more than just a powerful feeling; it is a tangible power that is self-supporting.

过去是，现在也是，它值得我每周出门捡拾垃圾。相比之下，我所获得的任何认可都是微不足道的。每当我发现自己没有做自己真正想做的事，我知道纪律就会回到我脑海中，我可以做我真正做的事并用我自己真正想采用的方式去处理。或许会有短暂的情绪让我一时偏离原计划的正道，但我却无须再受其支配。这不仅仅是一种强势的情感，这是一种可以自持的可视力量。

CHAPTER 8

It is OK to drop rubbish. Here's why.

I have come across various arguments why it is OK to drop rubbish. Here they are:

1. 'I am imposing my cultural values on people if I interfere with their long-established habits of dropping rubbish.' I heard this one from a young Canadian English teacher in China when he heard about my efforts and what I was doing. I didn't answer him at the time but I thought about it. The first thing was to ask if this was a cultural value. My students taught me this saying in Chinese and this is evidently well-known.

垃圾不落地，文明在手中。In Pinyin that is: lā jī bú luò dì, wén míng zài shǒu zhōng. In English my rough translation would be: If you don't drop rubbish on the ground, we can keep our culture intact and pure. It could also be: Don't drop rubbish; our culture is in your hands.

I asked a few Chinese people if they liked to have rubbish around them and of course nobody likes it. I would openly challenge anybody to find a culture anywhere in the world where the people really liked to have rubbish within their field of vision anywhere other than in rubbish bins or areas set aside for processing rubbish.

2. 'The rubbish breaks down quickly into the soil.' This is

第八章

乱扔垃圾有理的种种论调

我曾听闻各种乱扔垃圾有理的种种论调。比如说：

1.“如果我干涉他们长期形成的乱扔垃圾的习惯，那我就是把自己的文化价值强加于他们。”该论调出自一位在华任教的加拿大籍青年英语教师之口，他是在听到我的所作所为后有感而发的。对此，我当时没做回答但却思考过。我起初想问的问题是乱扔垃圾是否是一种文化价值。我的学生用汉语教会了我下面这句话，这显然是一句人所共知的话。

垃圾不落地，文明在手中。（其拼音为：lā jī bú luò dì，wén míng zài shǒu zhōng.）我用英语将其大致翻译为：你若不满地乱扔垃圾，我们便能使我们的文化完好而纯洁。原文亦可翻译为：切勿乱扔垃圾，我国的文化就在你的手中。

我曾向几位中国人讨教，问他们是否喜欢满地都是垃圾，答案自然是否定的。我可以公开挑战任何人，看谁能在世界任何一个地方找出一种文化，而该文化中的人真心喜欢满地都是垃圾，而不是让垃圾出现在垃圾箱或者处理场。

2.“垃圾会快速分解成泥土。”就餐巾纸或纸巾而言，这一论调

a common argument when used with paper napkins or paper tissues. It is also used when people drop things like orange peel, watermelon peel, apple cores or anything like this.

One of the most common bits of rubbish that is used once and then dropped in China is the paper towel. Other names this item may be called are tissue, paper tissue, toilet paper, paper napkin and so on. The item is the same and it does break down quickly. I don't know how long it takes precisely but in really rainy conditions with some foot traffic over it I am sure it could become part of the soil within a day or so. I don't know how many chemicals used to produce this item are good for the soil or bad for the soil, but I do know this. Until the paper breaks down and disappears to our eyesight, it is an eyesore. It does not break down quickly enough. In some places if you drop a paper tissue there may be 100 people who have to witness your laziness within a minute of your doing it. If those 100 people have come to see a beautiful scene or even if those 100 people are just on their way to work, I feel that it is still a selfish act and so my judgement of you is not very favourable. You could have put the paper into your pocket, into your bag, or into the rubbish bin that may be within 250 metres.

There is a further problem with this paper tissue that you have dropped. Rubbish attracts rubbish. I have noticed this phenomenon in different places. If I have some rubbish in my hand and there is a pile of rubbish waiting to be picked up by somebody I will add my little bit to the pile. Even on a pristine mountain path people do the same thing but often without thinking.

可谓司空见惯。随处乱扔果皮、瓜皮、苹果核的人也以此为由。

在中国,使用一次后便抛弃的最普遍的垃圾之一就是纸巾,又名卫生纸、餐巾纸,等等。这类东西性质相同,确实容易分解。我虽不清楚需要多长时间才能分解,但可以肯定在多雨时节,经人踩踏过后,纸巾倒有可能一两天便成为泥土的一部分。我不知生产这类东西需要用到多少种化学物品,或者对土壤有何利弊,但对如下事实我却是清楚的:纸巾未分解且从我们视线中消失之前,实在是有碍观瞻的。此物的分解并不快。在有的地方,你如果乱扔一团纸屑,之后不足一分钟,可能有 100 人不得不眼巴巴地见证你的懒惰。如果那 100 人是来观赏美景的,即便他们只是在上班途中路过,我觉得你的这一举动也不可谓不自私,因此,我对你的评价可不咋样。你本可以把那纸团放进自己的兜里、袋里或者可能仅有 250 米远的垃圾桶里。

你乱扔的那个纸团还会引发另一个问题:垃圾招来垃圾。我在不同的地方都注意到了这一现象。如果我手里有垃圾,而此时旁边就有一堆垃圾等人拾走,我会顺手将手头那点垃圾扔进那堆

If there is rubbish along the side of the path then there is no reason (in their unconscious minds) why they shouldn't add their rubbish to the general litter. If you drop a paper tissue then other people will drop other items that do not break down so quickly. I think everybody is aware that some plastics can take a very long time to break down and so I would suggest that dropping any rubbish anywhere is a bad idea.

3. 'By dropping rubbish we are creating employment for people who get paid to pick it up.' This is a very common argument and it is easy to defeat. I really think this is just an excuse for laziness.

If our councils or government have money to pay for people to pick up rubbish, then they and the general population would inevitably prefer that these people could be employed doing something that would benefit the community a lot more. There are always trees, flowers, and bushes to be planted and maintained. There is always regeneration work that can be done. There are always new skills that if they are learned can be put to use by governments and the general population. There will always be work that can be done and if we don't have to spend our money on people to pick up rubbish then we can spend that money on far better things.

Another important point needs to be made about this argument and that is that if you say it is OK to drop rubbish you will never be able to employ enough people to pick it all up. Why is this so? It

垃圾里。即便是在原始的山间小道上，人们也常常不假思索地这样做。如果沿小道有垃圾，人们（潜意识里认为）没理由不把自己手里的垃圾扔进去。如果你扔一团纸巾，那么其他人就会扔其他不易分解的东西。我认为人人都意识到了有些塑料品需要很长时间才能分解，因此，我想说的是，随地乱扔垃圾是不可取的。

3.“我们扔垃圾，是替受雇收拾垃圾的人创造就业机会。”这一论点十分普遍，但是经不起推敲。我真的认为这不过是懒人的借口而已。

如果我们的服务机构或政府有钱雇人收拾垃圾，那毫无疑问的是，他们与大众无疑更情愿雇请这些人做一些更有利于社区的事情。花草树木总是需要人栽种、维护的。总有修复工作需要人去做。总有一些新的技术需要掌握，而后被政府和大众采用。总有工作可做，如果我们不必花钱雇人收拾垃圾，那我们可以把钱花在更好的事情上。

关于这种论调，还有一个需要弄明白的要点，那就是，如果你认为可以随地乱扔垃圾，那么收拾垃圾的人是永远雇不过来的。这

is clearly logical and I don't think it needs to be explained. Maybe I can't explain it because it is so simple but I will try. If it is OK to drop rubbish then it is OK for the people you are paying to pick up your rubbish to drop their rubbish, so you must pay more people to pick up their rubbish and so this is a never-ending cycle. I come back to my initial point and that is that this argument is an excuse for sheer, naked laziness. There will always be a bit of rubbish that needs to be manually picked up because of the forces of nature and the occasional mistakes that people make. That is all there should be in my opinion.

从何说起？这在逻辑上一眼即明，我想无须解释。也许正是因为太简单，我才解释不了，不过，我可以试试。如果可以随地乱扔垃圾，那花钱请人员来清理你扔下的垃圾也是可以的，但你得花钱请更多的人来清理垃圾，这可是个永无终点的循环啊！回到起始点，那就是，此处的论调不过是赤裸裸的懒惰的遮羞布而已。由于自然的力量以及人们偶犯的差错，总会有些许垃圾需要人去捡取。这就是我的全部看法。

CHAPTER 9

What can I do?

I have long ago lost count of the number of people who have asked me this question when we are talking about environmental concerns. In this chapter I am going to talk about two ways of answering this question. Both of them start and finish with the premise that you should do what you can do.

We all have different abilities. It doesn't matter whether you are talking about physical, intellectual, mental, emotional, spiritual, social, leadership, managerial, teaching, or any other type of ability: We all have different levels and combinations of these abilities. The big thing for many people and I hope it is important for you too, is how we answer the following question: Are we using the abilities we have? This question for me automatically leads to questions about whether or not we are developing the abilities that are latent within us but that is a digression that is not the purpose of this chapter. The purpose is to point to the importance of using the abilities we have and allowing other people to use the abilities they have.

If you are a scientist or a chemical engineer, then you have skills that may lead to better products that do not pollute. You may find solutions to any one or more of the myriad of problems that combine to give us the pollution problems we have. I can't

第九章

我能做些什么

有多少人向我问起环保问题，我早已记不清楚了。在本章中，我打算讨论应对环保问题的两种方法。两种方法均由“你应该尽力而为”作为前提和落脚点。

我们都拥有不同的能力，无论你谈及的是体力、智力、精神、情感、灵性、社交能力、领导力、管理能力、教学能力，或是任何其他种类的能力：其中几种我们在不同程度上都具备。我希望，多数人的大事也是你认为重要的事，那就是我们该如何回答下面这一问题：我们都在运用自己拥有的这些能力吗？就我而言，这个问题自然而然地引出了一些有关我们是否在拓展自己潜藏能力的问题，不过，这个题外话不是本章的目的。本章旨在表明运用我们的能力以及让他人运用他们能力的重要性。

如果你是一名科学家或化学工程师，那么你可能就有能力生产不产生污染的产品，你还有可能找到方法，去解决任何一种或多种给我们造成污染困扰的问题。这一点我无力做到，只有你能，我

do that. Only you can and I am depending on you.

If you are a politician then you have the capability to promote practices or laws that can change the picture of the environmental future of your country (or state or city or town). From that beginning, you may help change our world's future. I can't do that. Only you can and I am depending on you.

If you work in your home to care for your family you have the capability to buy wisely, to cook and clean wisely. To do that you have to educate yourself about what impacts our world environment in a bad or good way and from there you can buy, cook, clean and care wisely. I can't do that. Only you can and I am depending on you.

It doesn't matter who we are or what we are doing. It doesn't matter if we are students, farmers, prisoners, wealthy or poor. It simply doesn't matter when it comes to the question of what we as individuals can do about our environment. We can all do a little or a lot according to our abilities. What we can do, we should do. If you are waiting until your neighbour does the right thing before you do it, then you are abrogating your responsibility to somebody else and it does not belong to them. Your responsibility is yours and yours alone. Don't wait to do the right thing either as an individual or a corporation or even as a country. Right now I am not being very humble about giving my opinion but I do realise that not everybody will agree with me. If you don't agree with me then I am more than willing to listen to your point of view and even change my own because you are thinking about it. You have an opinion. Now we must reach out together in whatever

就仰仗你了。

如果你是一位政治家，那你就有能力改进一些习惯做法或修改法律，这些均可让你所在的国家（州、市或镇）的环境面貌焕然一新。长此以往，你有可能改变我们这个世界的未来。这一点我无力做到，只有你能，我就仰仗你了。

如果你在家中照顾自己的家人，那你就有能力在购物、做饭、洗涮方面做出明智的选择。要做到这一点，你得了解什么对我们这个世界的环境是有益的，什么又是有害的，进而明智地购物、做饭、洗涮且照顾家人。这一点我无力做到，只有你能，我就仰仗你了。

我们是谁或是在从事什么工作并不重要，不管我们是学生、农民、囚犯、富人，还是穷人都不重要。作为个人的我们，仅就可为环保做点什么而言，上述问题都不重要。根据能力，我们大家或多或少都可以做一些事。我们应尽力地去做。如果你要等到自己的邻居做对了才肯去做，那你是在把你自己的责任推给不该负责的人。你的责任就是你的，也只能是你的。做正确的事不应以个人、公司或者国家的身份来区分。此刻，我意识到在提出个人的想法时有些不够谦逊，也意识到并非人人都同意我的观点。如果你不同意我的观点，那我非常乐意倾听你的观点，甚至愿意因为你在思考同

way we can to reach the people who have never thought about it. We must find a way to help people realise that breathable air, drinkable water, and having food to eat that will not do us damage may be important. Maybe these things are actually more important than a new cell phone or new car, or new computer. Maybe these things are actually more important than idly dropping a cigarette butt on the ground. Maybe these things are more important than using too much detergent to wash the dishes in our home. Maybe you and I should stop spending so much money on our ability to inflict war on others (or defend ourselves from others) and start spending these vast resources in ways that will actually benefit us all. Maybe—maybe! Find your own maybe and turn it into a 'Yes! What I do is important!'

The second way of answering this question has a million different answers and it does not depend on any maybe. It depends on you and I being willing to stand up and take our actions in full view of the world with as much consciousness and honesty as we can. If we are doing something that we know is not the best thing for the environment then we have to explain our rationale for doing so.

We can make a very simple example first of all. (I have a cigarette and then I drop the butt on the ground. My rationale might be that it would burn a hole in my pocket if I put it there.) I am sorry but that is not a very good excuse for dropping rubbish on the ground. You can put out the fire on your cigarette and then put it in your pocket so that you can dispose of it properly later. Maybe you would explain yourself by saying that one cigarette

样的问题而改变自己的看法。毕竟你有你的想法吗。这下，我们必须团结一致，想方设法让未曾思考过这一问题的人们了解我们的想法。我们必须想办法让人们意识到，可自由呼吸的空气、可饮用的水和不伤身体的食品的重要性。事实上，这些东西也许比一部新手机、一辆新车或者一台新电脑更重要，也许比漫不经心随地扔个烟头更重要，也许比在家里洗盘子时用过多的清洁剂更重要，也许你我应该停止耗资去向他人发动战争（或保护自己不受他人伤害），转而用实际可造福世人的方式来使用这巨大的财富。也许，也许！尽力发现你自己的“也许”，并将其转化为一种“没错！我做的事情很重要”吧。

回答该问题的第二种方式能带出一百万个答案，该方式并不依赖任何的“也许”。它依赖你我是否愿意以足够的意识和诚实的态度，当着世人的面站出来，并采取行动。如果我们明知自己正在做对环境不利的事，那我们就得为自己的所作所为给出一个合理的解释。

我们不妨先举个简单的例子。假设我抽了一支烟，然后将烟头扔到地上。我的理由是，如果将它放进兜里，那它有可能把我的兜烧个洞。很抱歉，这并不是一个随地扔垃圾的好借口。你可以灭掉烟头，然后将其放入兜里，以便方便的时候处理。也许你会辩

butt is not going to cause a problem. I am sorry but when half a billion people drop several butts every day, your argument falls apart. It comes down to the basic fact that every little cigarette butt adds up to being a mighty big problem. Almost everybody is aware that the various chemicals in most cigarettes are very poisonous to our bodies and the earth is not very different to our body. Our existence is dependent first of all on our physical body but then also on our earth. Neither can be poisoned with impunity. Now comes my big point on this matter. I must be willing to go and pick up the butt that is carelessly cast aside and then, if I can, ask the person who threw it to not throw any more away. Can I do that? Can you do that? I don't know yet. I hope we can.

A less simple example could be when we buy a new car. It could be a new bicycle, a new computer, a new mobile phone or a hundred other such consumable items. We need to ask ourselves the question: Do we really need such an item? Too often I am afraid that the answer is 'No, we don't.' The problem with buying new gadgets or toys or appliances is that the old ones have to go somewhere and most of the time, most of them end up as scrap or some form of rubbish. Too often there are parts of these items that then leach a cocktail of pollutants into the earth. The big issue is whether or not we would be willing to ask our friends the same type of questions instead of simply admiring their latest and greatest new acquisition. Please understand that I am not coming from a 'holier than thou' viewpoint here. I love new toys. I want another bigger, better mobile phone than the

解，说一个烟头是不会引起麻烦的。很抱歉，如果五亿人每天都扔几个烟头，那你的论调便不攻自破了。归结起来，事实是，每个小小的烟头加在一起就构成了一个巨大的问题。几乎人人皆知的是，多数香烟里的各种化学物质对我们的身体极其有害，而地球和我们的身体没多大不同。我们的生存首先依靠我们的身体，其次靠我们这个地球。两者都经受不起污染。现在说说我对此事最主要的看法：我一定愿意去捡起那个被随意扔掉的烟头，然后，如果可能的话，恳请那位扔烟头的人别再随意乱扔。我能那么做吗？你能那么做吗？我还不知道，真希望我们都能那么做。

我们不妨再举一个不那么简单的例子，即当我们买一辆新车，或一辆新自行车，或一台新电脑，或一部新手机，或一百个其他类似的消费品时，我们需要扪心自问：我们真的非要这东西不可？多半情况下，我恐怕人们的回答是"不，我们不是非要不可"。购买新产品、玩具或者电器带来的问题是，那些不用的旧物件总得有个去处，而它们多半沦为废弃物或者某种形式的垃圾。这些物件的零部件中的一些混合污染物会泄漏并渗入土壤中。关键问题是，我们是否愿意向我们的朋友们提出同样的问题，而不是简单地欣赏他们那最时新、最了不起的购置品。请务必理解，我此刻并非站在道德至高点上指手画脚。我爱新玩具。我想要一部比现在这部更

one I have. I need a new mobile phone. My computer is almost four years old now. Surely I am due for a new one soon. I really want one of those really fast bicycles I saw when I was last in Australia.

And so the story goes on. What can you do? What can I do? We can pick up rubbish. We can be more thoughtful about what we are buying and consuming. We can lobby for social and legal changes to make our physical world a better place. We can choose our holiday destinations carefully. We can spend our money with more care and wisdom. Above all, we must do whatever it is that we can do and then we must keep on doing it.

大、更高档的手机。我需要一部新手机。我那电脑差不多用了四年了。我购置一台新电脑的时刻显然很快就到了。我上次在澳大利亚见到过一些骑行速度相当快的自行车，真想购买一辆！

故事可以不停地讲下去。你能做点什么呢？我能做点什么呢？我们能捡拾垃圾。我们能对自己购买以及消费的东西多一些思考。为使我们生活的世界更加美丽，我们能为要求社会及法律的变革而奔走呼吁。我们能仔细选择我们的度假地。我们能更细心、更理智地进行消费。更重要的是，我们做事必须竭尽所能，并持之以恒。

CHAPTER 10

What are you?

One day I was talking to another Westerner while I was in China and he asked me if I was the guy who picked up rubbish. When I answered that I was, he said, 'Oh, you are an environmentalist.'

This was said in such a way that it sounded very much like a derogatory accusation. I was taken aback and my first thought was to angrily retort, 'Well. Who isn't, you idiot?' Thankfully I managed to hold my tongue at that moment but it set me to thinking some more.

Who among the whole of human population likes to walk amongst rubbish if they had an alternative? If you had a choice between walking in mountains where there were trees, bushes, rocks, flowers, and streams or walking in mountains where there was rubbish, trees, litter, rocks, refuse, flowers, and food packaging cast by the wayside: What would you choose?

I got a little bit angry when I was accused of being an environmentalist. It epitomised to me the blindness with which some people approach their world. They do not see the rubbish. The do not smell the air. They seem to taste their food but I wonder if this is just for show. I wonder if they are feeling or tasting anything like I experience. They seem to love their families but I wonder if this is not just dependency. I know that if you ask them

第十章

你是谁

在中国期间，我有一天正跟另一名西方人交谈，他问我是不是那个捡拾垃圾的人。我回答说正是本人，他说道："哦，你是个环保主义分子。"

此人说话的方式，让人怎么听怎么像是在损人与责难。吃惊之余，我起初的想法就是愤怒地回敬他一句："是啊，谁不是呢？你这白痴！"多亏我当时控制住了自己，不过，此事还真让我思考良多。

整个人类中，如果有别的选择，有谁会愿意行走在垃圾之中呢？如果让你选择是行走在有树木、灌木丛、岩石、花卉、溪流的山中，还是行走在有垃圾、树木、丢弃物、岩石、垃圾堆、花卉及路边弃置有食品包装袋的山中，你会怎么选呢？

当我被指责是一名环保主义分子时，我有些恼怒。在我看来，这些字眼概括出了某些人为人处事的盲目性。他们对垃圾视而不见，他们也闻不出空气中的味道。他们似乎吃东西吃得很香，可我却在想他们是不是在装模作样。我不知他们对任何东西的触觉和味觉跟我所经历的是否一样。他们似乎都爱自己的家人，可我却

to walk down a street littered with rubbish and the cast-offs of the modern society including the fumes, or a scenic garden trail with trees, flowers and clean air, they will choose the latter. I hope that they would choose the latter. Maybe I really am a crazy idealist. Maybe they like the pollution, the litter, the noise, and the fumes. Maybe these assaults on the senses are what now pass for excitement or normality. Maybe they are not alive in the sense that I feel I am alive.

Who amongst us likes to swim in an ocean skimmed with oil and the oil soaked carcasses of dead and dying birds and sea creatures? How about a relatively clean ocean that is simply discoloured by the effluent from a local factory? No one does. That is who. Are all these people environmentalists?

Who likes to walk down a street while breathing fumes that they know are damaging their lungs to the point that they know it is damaging their health and prospects for a healthy life? Who would want to exist in this environment if there was an alternative for them? No one would willingly submit to this if they could avoid it. Are all these people environmentalists?

As you can see I was riddled with self-doubt about this subject so I did a small survey and asked about six people what they thought about it. At first they all looked at me as if I was stupid. The answer to all these questions was as self-evident to them as it was to me.

If we all would really prefer that we inhabit a pristine or 'nice' environment then we must ask different questions of ourselves and our neighbours. We must start acting in accordance with our

不知这是否仅因依赖性所致。但我知道，如果你问他们是愿意沿着满是垃圾和现代社会废弃物（包括各种废气在内）的街头行走，还是愿意沿着满是树木、花草和清新空气的观景花园小道行走，他们会选择后者。我希望他们会选择后者。也许我真是个疯狂的理想主义者。也许他们喜欢污染、纸屑、噪音和废气。也许上述种种对五官的冲击便是现代人所谓的刺激和常态吧。也许他们并不活在我所认为的活着的意义中。

当一片海域漂浮着油污，已死或行将死去的鸟儿及海中动物的尸体浸泡其中时，我们当中有谁愿意下去游泳？那么一片仅因为附近工厂排放污染物而变色的相对干净的海域呢？两者的答案都是否定的。这就对了。难道这些人都是环保主义分子吗？

明知吸入废气会损害自己的肺，进而伤害身体，无法继续健康地活下去，有谁还愿意沿街而行呢？如果还有别的选择，有谁会愿意生活在这种环境中呢？如果能够避免，谁都不会心甘情愿地屈从。难道这些人也都是环保主义分子吗？

如诸位所见，就这一主题，我深受自我疑虑的困惑，于是进行了一次小型调查，向六个人问及对此事的看法。起初，他们全像看傻子似的看着我，对所有问题的回答，对他们、对我本人都是不言自明的。

如果我们都真心地希望自己居住在一种原生态的或者说“不错的”环境中，那就必须问我们自己和我们的邻居不同的问题。我

deeper desires. We must start acting as if there will be a tomorrow.

I am going to leave that up to you now. After all, only you can answer the questions:

What are you?

Do you believe there will be a tomorrow?

Do you care about what we leave to the younger generations? If you are a part of a younger generation, do you care about anything?

Do you believe that you do make a difference?

们必须遵从我们内心更深层的渴望，行动起来。我们必须带着“会有明天”的理想行动起来。

现在，我打算将这一行动的机会留给诸位。毕竟只有你们才能回答这些问题：

你是谁？

你相信会有明天吗？

你在乎自己给后人留下什么吗？如果你就是下一代人中的一员，你在乎什么？

你相信自己真能有所作为吗？

CHAPTER 11

Success

The reason I am including this chapter is because it might help explain how I have kept at the job of picking up rubbish in a difficult situation. Let me first explain just how difficult the job is.

The amount of rubbish is massive and I am just talking about one little mountain in one little province of a very large country. Initially even the amount of rubbish on the walking area of the path was mind-boggling. If you dared to look to the side of the path, then the amount increased a thousand fold. Once you stepped off the path to pick up the rubbish you could see the amount increased again by a similar factor. Then there were/are the setbacks that happen. With a great deal of effort we would manage to clean up an area only to come back a week later and find that people having a BBQ party had left a mess as bad as what we had previously cleaned up or even worse than it had been before. Another similar thing often happened. Sometimes over many weeks we would successfully clean a kilometre length of track only to come back after a holiday when there was fine weather to find vast quantities of litter along the whole way. There may have been thousands of tourists along that track during our absence and many of them would leave us a little present. At

第十一章

成　功

我将本章内容写进书里，是因为这可能有助于解释我为何在逆境中能一直坚持捡拾垃圾这项工作。先让我解释一下这项工作有多难吧。

垃圾之量十分巨大，而我谈及的仅仅是一个泱泱大国中一个区区省份里的一座小山而已。起初，就连小道上步行区域内的垃圾数量都令人震惊。倘若你敢朝小道边上望去，那垃圾之量可以增加一千倍！一旦你离开小道去捡拾垃圾，垃圾的数量又会增加约一千倍！随之而来的便是发生过的种种令人败兴之事。我们花了不少精力才清理干净一片区域，可等我们一周后回到原地，搞过烧烤派对的人又把那地方打回原形，甚至不如从前。类似情形常常发生。有时，历经数周的努力才清理完的一公里长的小道，赶上好天气的假期一过，我们回来时发现沿路又布满了数不清的丢弃物。我们不在的期间，有数以千计的游客路过那条小道，其中不少人都给我们留下了“小礼物”。在引进一些绿色环保的大型垃圾桶

one point after the introduction of some big environmentally green looking bins, we found that after the bins starting looking full, someone was helpfully emptying them—onto the ground next to the bins.

Some of the rubbish we have cleaned up had been accumulating over many years. We would clean a little only to find that a vast reservoir of litter lay just beneath. Rarely did I allow any short cuts in our efforts to clean these ugly scars on the landscape. We would simply pick it up and keep picking it up until it was all clean.

After all that, one must remember that the physical effort involved in just walking up the mountain is difficult for many people. When you add in the extra effort of bending down and scrabbling around with rubbish and rubbish bags it can add up to be a physically demanding job.

I recently watched one of the news video clips of me on the mountain and I noticed that I said that picking up just one bit of rubbish meant to me I was successful that day. I realised then that would be a strange idea to many people. For about 20 years now I have been studying different aspects of success and this incident was a manifestation of something I learned from probably several different teachers.

I once heard an American motivational speaker by the name of Skip Ross. I listened to tapes of him speaking many times and I'm sure I learnt a lot but now this is the one thing I remember well from him. It is a definition of success: 'Success is the progressive realisation of a worthwhile dream.' It is easy for me

后的一段时间里，我们发现，当那些垃圾桶看上去快盛满时，就有“好心”人想清空它们，也就是将垃圾倒在那些垃圾箱附近的地面上。

有些我们清理掉的垃圾已经堆积了许多年。我们在清理一点后发现就在下面竟然还有一个巨大的废物库。在努力清理有碍观瞻的疤痕的过程中，我不大允许自己走什么捷径。我们总是一遍遍地清理，直至完全清干净为止。

爬过山的人一定知道，多数人光上山就已经气喘吁吁。当你带着一个个装着垃圾的垃圾袋，弯腰捡拾、爬上爬下，这些加在一起对体力无疑是个严峻的挑战。

我最近观看了我在山上的一个新闻片段，且注意到自己曾说过，就算只捡起一片垃圾，我那天也算是成功了。之后，我意识到这话在很多人看来是个奇怪的观点。近二十年来，我一直在研究成功的不同层面，这个观点正体现我从来自不同层面的几位老师那里得到的收获。

我曾听说过一位名叫斯基普·罗斯的美国励志演说家。我多次聆听他演说的磁带，并确信自己学到了许多东西，其中我记得非常真切的便是如下一事，即他对成功的释义：“成功就是对值得追寻的梦想的逐步实现。”作为一名澳大利亚人，我轻而易举地记住

to remember Skip Ross because to my Australian ears he had the most annoying American accent but as it turns out, this has meant that I have never forgotten him or that quotation. The way I applied this definition of success to picking up rubbish may be obvious to you. The dream is and has always been, having this mountain free of rubbish. If we are making progress toward that dream then we are successful.

Somewhere else I learnt that we can set our own measures of success. One of the advantages of this idea is that anybody can then set themselves up to be successful. If you are successful you can then feel good about that success and you are then much more likely to go on to achieve further success. Too many people fail at something and then they continually fail and then they give up. Most people know that road and mostly, it doesn't work very well. You can set your bar low or you can set it high but in the short term it is a pretty good idea to set the bar a little low until you get success firmly ingrained into your psyche. You can get to be so used to being successful that you expect it and then you can go on to demand it of yourself with harder and higher goals than you ever thought possible.

The way this has worked for me with picking up rubbish has been that I set the bar ridiculously low. If I picked up one bit of rubbish I was successful. This has meant that I am always successful every time I go out. Don't forget that sometimes I go out and my back hurts, my foot hurts, I don't feel well, I have the flu, I'm very tired, it's raining, it's snowing, it's very cold, it's very hot, and so on. Also don't forget about those huge setbacks

斯基普·罗斯,是因为他讲话带有那种让人最为讨厌的美国腔,这也意味着我从没有忘掉过他或者他说的那句话。我如何将这一成功的释义运用于捡拾垃圾,在你看来也许不言自明了吧。至于梦想,那就是,也一直是,让那座山不留半点垃圾。如果我们朝着这一梦想不断迈进,我们就是成功的。

在别的地方,我还学到了我们可以自行设定成功标准。这一想法的益处之一就在于,任何人都能借此获得成功。一旦你成功,你会因成功而感觉心情不错,之后更有可能继续取得更大的成功。因某事失败而接二连三地失败,最终放弃的人太多太多!多数人对这一方法都了如指掌,但尝试后的效果往往不佳。你可以抬高或者放低自己的标准,但就近期而言,降低标准不失为一个好主意,直到你对成功有深刻的领悟为止。你会逐步地适应成功之感,最终期待它,继而用超出自己预想的日渐苛刻的种种目标来要求你自己。

我捡拾垃圾之所以奏效,是因为我把自己的成功标杆定得低到可笑的程度:如果我捡起一片垃圾,我就算成功了。这就意味着我每出门一趟,我总是成功而归。千万别忘了,我有时出门时会腰疼、脚疼、身体不爽、患流感、疲惫不堪、遇着雨天、赶上下雪、天气

that have often totally thrown out my little dreams for the day.

Almost every time I set out to spend my allotted time in picking up rubbish I have a goal to clean a certain area or to maintain a certain length of track. Often I have had to adjust my expectations and rarely have I exceeded them. When I have a fallback position that means I have been successful regardless of my goals or dreams for the day, I can still be happy and keep on working.

Now we can start thinking about persistence and how that fits into our picture of success. In my studies about success I have read and listened to many people including Bob Proctor, Dennis Waitley, Bill Harris, Ken Wilbur, Tony Robbins, and many more great teachers. Without exception, I am quite sure that persistence is one of the things they all talk about to some degree or another. It is one of the keys to my success in everything I have ever been good at. Personally I think my persistence is closely related to my ability to be obstinate. This is not such a great quality to be bragging about but the fact is that I have sometimes been very set in my thinking and subsequently very slow to learn.

One thing I have not been slow to do though is to dream great dreams. I don't think there is anything wrong with this except that if you dare to dream great dreams, you are going to be disappointed sometimes. You are going to fail sometimes if you set great goals. Is there anything wrong with that? I see a lot of people afraid to dream and afraid to set goals and I think the reason is that they are afraid to fail. They are afraid to be disappointed.

寒冷、酷热难耐等，也别忘了那些巨大的挫败，它们常常将我每天的小梦想砸得粉碎。

几乎每次在我分配时间捡拾垃圾的时候，我都会制订清理一定区域或一定距离的目标。我常常不得不调整自己的期望指数，很少有超过那些预期目标的情况。当我拥有供选目标时，那就意味着我已经成功，且不管当天有什么目标或者梦想，我仍然是快乐的，并继续工作下去。

接下来，让我们来思考毅力及其如何跟成功相容吧。在对成功的无数次研究中，我曾阅读且聆听过不少大师的教诲，他们是鲍勃·普罗克特、丹尼斯·魏特利、比尔·哈里斯、肯恩·威尔伯、托尼·罗宾斯等。我确信，他们在某种程度上都毫不例外地提到过成功的要素之一就是毅力。在我擅长做的那些事情中，毅力便是我成功的钥匙之一。我个人认为，我的毅力与固执己见密切相关。这虽不是什么值得炫耀的美德，但事实是，我的思维一直较为固定，结果导致学习十分缓慢。

但是，有一样事情我做起来并不迟缓，那就是展开梦想的翅膀。我认为这没什么不妥，只是你若敢展开梦想的翅膀，你有时就会失望。如果你树立远大的目标，你有时会失败。即便那样，又何错之有呢？我见过不少人害怕梦想，害怕树立目标，我认为其原因是他们害怕失败，害怕失望。

Sometimes we have to face up to the tough stuff and this is tough. Here it is! If this is you then it is time to develop a bit of back bone. It is time to grow up and dip into the deep well of courage that is there within you. The failure of never trying, never dreaming, and never setting worthwhile goals means to me that you have never fully lived. Let me tell you a couple of my big dreams.

I dream of China leading the world in all things to do with environmental protection. These things would include all the technological and social products that need a lot more development and implementation. One of the big hurdles for any government and its people is that often people look to government to do all the work. They look to government for the leadership. They look to their government to show them and educate them what to do. That's all well and good but I think that it is up to each and every one of us to do the work, to be leaders and to educate ourselves. To be successful in looking after our environment and improving it takes effort from everybody but it starts with me. It starts with you. Sure it also starts with the government but they are already doing a lot. We can complain that they are not doing enough but what are we doing? What am I doing today? What did you do yesterday and what will you be doing tomorrow? You have to ask these questions as well.

Dreams! Dreams! Dreams! I would be happy to dream of Australia, my homeland leading the world in environmental protection but for now at least, the people and government of Australia seem to be heading in the opposite direction. I really

我们有时不得不迎难而上，可要那么做也十分艰难，原因就在这里！你若就是这样的人，那现在就是你直起脊梁的时候，是你长大成人且挖掘隐藏在你内心深处的勇气的时候。在我看来，因从未尝试过、从未梦想过、从未树立过有益目标等而导致失败，这意味着你从未充实地生活过。且听我说说我的几个远大梦想吧。

我梦想中国的环保工作在各个领域全面引领全世界。这包括所有需要大力开发并推广的技术及社会产品。对任何政府及其人民来说，巨大的障碍之一就是，人们期待政府将所有工作来个大包大揽。他们指望政府牵头，他们期望政府做出示范，教他们做。这个想法无可厚非，但我认为工作还得由我们每个人自己去做。要在环境保护方面取得成功并给予改善，这需要人人努力，不过应从我、从你开始。当然也应从政府开始，但政府已经做了不少努力。我们可能会抱怨政府做得不够，可我们又在做什么呢？我自己今天在干什么呢？你昨天做了什么？明天你将会做什么呢？你也不妨自问一下。

梦想！梦想！梦想！我乐于梦想澳大利亚，我的祖国会在环保方面起带头作用，可惜澳大利亚人民和政府，至少目前，像是在朝反方向行进。我真的希望这种情形会改变，也许我将回到澳大

hope that situation changes and perhaps I will return to Australia and try to be a part of that change. The idea of being involved politically fills me with dread. Politics in Australia seems to me to be a pretty dirty business. There is so much back stabbing, parochialism, and people only looking at things with one eye. Also, many Australians do not look very far into the future and this short-sightedness is reflected in the action of the government. We used to call ourselves 'The Lucky Country' but I think we have wasted away so much of our luck. My home state of Tasmania is still a beacon of environmental care and cleanliness but even there, that reputation is always under threat from the seeming necessity to have more progress, more wealth, more people, more work, more exports, more houses, more tourists, longer lives, and so on. I will never be against progress but if that progress comes at a cost to the air we must breathe, the water we must drink, to the land that supports us, I will always be against it. This applies even if it means there will be less jobs or less money in the near future. Where will the future lead us? We can still dream of a good or bad future and therein lays a key to our actions—if we dare.

On a totally different tangent, another big dream for me is that I will run a marathon in under three hours by the time I am 70 years old. That is a pretty crazy personal dream but it is mine and I am not letting it go—yet. I believe it is still possible but right now I am not paying the price necessary to achieve this goal because I have other priorities. I am planning and scheming though and very soon I am going to start work on this dream. It is a crazy

利亚，努力成为改变现状的一员。一想到跟政治沾边我便满心恐惧。澳大利亚的政治在我看来是一件相当肮脏的事情，其中包含太多的暗箭伤人，狭隘的地方主义，人们看问题时总是以偏概全。再有，很多澳大利亚人看问题不会考虑太远，这种短视反映在政府的行为上。我们以前自称是“幸运的国度”，但我认为我们却浪费掉太多的好运。我的家乡塔斯马尼亚州仍然是关爱环境和整洁的代表，可即使在那里，因其需要更大发展、更多财富、更多人口、更多就业、更多外贸、更多住房、更多游客、更长寿命等，这种良好的口碑业已遭到威胁。我不反对进步与发展，但是如果这种进步与发展是以牺牲我们必须呼吸的空气、必须饮用的水源，以及维持我们生存的土地为代价，我将始终如一地持反对态度，即使这意味着不远的将来会出现就业机会或收入减少等现象。未来会将我们带向何处？我们可以勾勒一个美好或丑陋的未来，关键取决于我们的行动——就看我们有无胆识。

换个话题吧！我还有一个宏伟的梦想，那就是到我七十岁时，能在三小时内跑完一场马拉松。这是一个颇为荒唐的个人梦想，但这就是我的梦想，我还不想放弃。我相信这是可能的，不过目前我没打算为实现这一目标而付出必要的代价，因为我有其他要紧的事要做。尽管我还处在计划阶段，但我很快就会着手去实现梦

dream so I might not achieve it. That does not mean to me that I will automatically be a failure. Along the way towards that goal I will achieve a lot and that will be well worth the effort. I don't like failure in any form though so I will be working very hard to achieve my goal and in my mind's eye, it simply has to happen.

In the meantime I will be continuing to work on Da Luo Mountain while I am here in Wenzhou. My dream here is that you will be able to walk up the mountain and not see one bit of rubbish. I will run up and down this mountain and not see one bit of litter. Now that is absolutely crazy. After speaking to several people about this dream I think that I may be the only one who believes it can happen. Sometimes I don't even believe it can happen but then the persistence kicks in and I just keep working as if it can happen. Sometimes I feel like it is not my job to believe or not to believe. My job, like a soldier, is just to go and do the work that is laid out in front of me. My job is not to keep questioning. My job is to keep working until the job gets done or to die on the job. It's simple. I am committed. Full stop! Dream on! Wherever I am, the work must continue.

想。这一梦想很是疯狂，我有可能无法实现。但就我而言，那并不意味着我会自动认输。在那一目标的实现过程中，我会获益良多，而这正值得我付出如此多努力。当然，我并不喜欢任何形式的失败，因此，我将会为实现自己的梦想而努力工作，在我看来，一定会成功。

与此同时，只要我在温州，我将把温州大罗山上的工作干下去。我在这儿的梦想就是，你上山时看不见一点垃圾。我在山上来回跑动而不见一片垃圾。这可真是疯狂之极！在跟几个人谈及这一梦想之后，我想自己可能是唯一相信梦想会实现的人了。有时，甚至连我也不相信梦想会实现，不过，毅力的能量在内心涌动，于是我又不停地工作起来，好似梦想就要实现一般。我就像一名士兵，去做摆在自己面前的那项工作。我的工作不是不停地质疑，我的工作是不停地工作，直至完成或者自己以身殉职为止。就那么简单。我矢志不渝！别的少说！坚持梦想！无论我身在何处，工作必须继续。

CHAPTER 12

A Final Word

There are a multitude of problems facing humanity. In this chapter I want to put on my childish, probably naive point of view, and just say it as I see it. This is my book and I can write whatever I want but I do realise this is very self-indulgent. If you have read this far though, I hope you don't enjoy my musings in this chapter. If you don't already think I am crazy then you probably will soon.

In my lifetime, the population of our world has more than doubled. When we look at the history of the human race, this is a crazy statistic. What do we think we are doing? How do we think our little planet can support that type of crazy expansion? The trouble is we don't think. I mean the Chinese government was thinking ahead when they instituted the one child policy. That policy warranted some bad press in the west for a long time. I know this policy was not, and is not a 'pretty' thing for many people to cope with. It is an infringement on personal liberty and when people are forced to have abortions then it gets very ugly. I think that the alternatives are a lot uglier. If we allow people to produce an infinite number of children their future and the future for all humanity is much uglier than an abortion. I can sympathise with those people who categorise abortion

第十二章

最后的话语

人类面临着数不尽的问题。本章中，我打算陈述自己幼稚的甚至可能是天真的观点，并都是我真实的想法。本册子是我写的，因此可以写任何我想写的，但我的确也意识到，这样做显得太自我放纵了些。如果你已阅读到此，我希望你别欣赏我在本章中表达的沉思冥想。如果你认为我还不疯，那你也许很快就有那种感觉。

自我出生以来，这世界的人口已翻了一番多。当我们回顾人类历史，你会发现令你震惊的数据。我们认为自己都在干什么呢？我们凭什么认为这个小小的星球能承载那种疯狂的膨胀？问题在于我们自己不去想。我的意思是，中国政府预计到了这一点，于是才推出了独生子女政策。该政策长期招致西方的诟病。对许多人来说，我知道，这项政策过去、现在都是一件不被“看好”的事。那是对个人自由的侵犯，人们被迫堕胎成为一件非常难堪的事。我认为其他可选方案更让人难堪。如果我们让人们无限制地生育，那么他们以及整个人类的未来会比堕胎更让人难堪。我能体谅视堕胎为谋杀的那些人的心情，不过，当我们因缺乏远见而致使数以百万计的人群注定受贫穷、饥饿之苦，那种场面可比任何谋杀方式

as murder but when, by our short-sightedness we are dooming countless millions of people to poverty, hunger and the like, that situation is uglier than any murder can be. It should never come to that but given the way we have operated so far, it will probably continue to get that bad and worse. Why can't we force those people who can't think beyond their immediate interests and desires, to act for the greater good on a long-term basis? Why can't we as ordinary citizens hold our governments to account for not aiming for a zero population growth? The reasons that we don't are too often, sadly, because our own immediate self-interest is compromised if we do. In the meantime we as a species seem to be acting like a virus bent on self-destruction. I have never ever heard of a politician anywhere in the world campaigning on the basis of what we are leaving for our grand children or great grandchildren. There surely must be a politician somewhere in the world that is willing to make a stand for their children and grandchildren. If not—why not? Are we all so disinterested in future generations and the planet they will have to live on?

Another crazy thing that just amazes me is nuclear power stations. I am only thinking about one aspect here and that is the waste. For the sake of power for our houses and our industries now, we are willing to leave behind an awful pollution problem that lasts for hundreds of generations of our descendents. What do we think we are doing? Have we such little respect for our children and their offspring that we are willing to give them this type of heritage. Don't try and tell me we can bury the waste or keep it safe because that has been proven to be wrong, wrong,

更难堪。那种场景绝不该出现，可按截至目前我们的运作方式来看，这事极可能会持续糟糕下去。从长远的角度看，对那些只考虑眼前利益和欲望之人，我们干吗不强制其为更大的利益而行动起来呢？作为普通公民，我们怎么不让自己的政府为不制定人口零增长目标给个说法？可悲的是，我们不这样做的种种原因往往是，如果我们那样做的话，我们的眼前利益就会打折扣。与此同时，作为一个种族，我们的行为让我们就像执意要自我毁灭的一种细菌。在世界任何一个角落里，我从未听说过某位政治家是为子孙后代的利益而参选的。在世界某个角落里，一定有这么一位政治家愿意为子孙后代表明某种立场。如果不——为什么不呢？我们对后辈们以及他们生活的星球真的如此冷漠吗？

让我惊讶的另一件疯狂之事就是核电站。我在此想到的只是其中一个方面，那就是废料。现在仅为了自己的家里和工业所需的能源，我们竟愿意留下一个可怕的污染问题——一个会持续影响数百代人的问题。我们可曾想过我们到底在干什么吗？难道我们对自己的孩子及其子孙就剩下这么一点尊重，以至于心安理得地给他们留下这种遗产？千万别试图告诉我，说什么我们可以埋掉废料或将其妥善存储起来，因为这一谎言已被证明是错、错、错！

wrong and just wrong. Let's keep it ultra simple. Our planet is not just a lump of rock that never changes. It is changing all the time and there are quick historical changes just as there are slow changes such as when the continents move. The quick changes I am thinking about are caused by volcanoes, earthquakes, extreme climatic events, and meteorites but I am sure there would be others. There is no safe place on Earth to dump nuclear waste. What incredibly selfish gene is so dominant in us that we even think of building another nuclear power plant facility anywhere on our tiny planet. If you were to ask me on a bad day, I would tell you that we should drop a nuclear bomb on any and all of these power plants. For the sake of the long-term health of our planet, we would be better off. That is a ridiculous idea so I apologise.

Why do we allow our cities to grow to the point where they are unhealthy places to breathe? What is that about? Air is our most basic requirement for life and we allow this to happen in the name of economic necessity. I will allow that it seems that we can survive in an increasingly hostile environment. Somehow we are adapting to being able to live in and breathe in an air environment that does kill us if we are a bit vulnerable, and it does shorten our life expectancy, but if we are earning enough money to eat, procreate, and live for a while, I guess we are happy enough as a species. The trouble is that if we continue on this path, we get to the point where we are buying bottled air just to live. Already we are buying bottled water in many places and starting to think that this is normal. Water is only the second

大错特错！这件事情再简单不过了。我们居住的这个星球，它不只是一块不曾变化的岩石而已。这个星球一直都在发生改变，恰如随着各个大陆漂移而出现缓慢变化一样，也会出现种种快速的历史变化。我正思考种种快速变化是由火山、地震、极端气候事件，以及陨石等引起的，可我确信还会有其他因素。地球上没有安全的地方可填埋核废料。是我们身上的那令人难以置信的自私基因致使我们还想在我们这个小小星球上再建核电站设施。你若在我心情不爽之时问起我这个问题，我会告诉你，我们应该在所有这些核电站的上空扔下一颗原子弹。替自己生活的这颗星球的长期健康着想，我认为这样做会让我们生活得更好一些。这是个荒谬至极的想法，因此我恳请原谅。

我们干吗非让自己的城市变成臭气熏天之地不可呢？这都是怎么搞的？空气可是我们生存最基本的需要啊！而我们竟以经济发展需要的名义让这一切发生。如果我们能在日渐恶劣的环境里生存下去的话，我似乎觉得还可以容忍。不知咋的，在一种可置虚弱者于死地的空气环境中，我们竟能逐渐适应生活和呼吸，但这的的确确会缩短我们的寿命，不过，如果我们仅为了挣钱吃饭、繁衍后代、生活一段时间，我猜我们作为一个物种倒也是其乐融融。麻烦的是，如果我们这样继续下去，我们会弄到靠买瓶装空气才能活下去的地步。我们很多地方的人已经在买瓶装水了，还认为这没什么不正常的。对我们生存最为重要的东西中，水只排在第二位。

most important thing to our survival. When we get to the point that even the rain is poisonous to us, we have a sad state of affairs. This already happens in some places in the world. Do we have to wait for a bloody revolution to fix these problems?

The message of this book is that we can change anything and everything about our environment. The only problem is that we have to do it one step at a time. Our problems were created one little bit at a time and maybe we have to solve them one little bit at a time.

Pollution can be cleaned up. First it must be done at the source but as in the cleaning up of litter this can be done concurrently with cleaning up the result. Habits of littering can be changed easily by having the people who do the littering either inspired enough to start picking up litter or we could compel them to do so for a short while. Maybe other problems are not solved so easily but that doesn't have to mean that they can't be solved.

I have dreamt of every building, every factory, and every vehicle being powered by its own non-polluting power source. How is this to be done? I don't know exactly but the solution may be in the form of a small or larger device scaled up to the requirements of a given situation. It might be some sort of torque generator that produces hundreds or thousands more units of power than that required to start it running. I know that such a device is not only possible but that it has already been created. Don't tell me this is not possible because I know it already exists. There is just the inertia of disbelief, big money interests, and lack of government will that is stopping it.

等我们到了连雨水都有害身体健康的那般田地，那我们的处境就悲惨了。这一现象，在世上某些地方已经出现了。我们非得等到发动一场血腥革命才去解决这些问题吗？

本册子欲传达的信息是，对于我们的环境，我们是可以改变一切的。唯一的问题是，我们只能一步一步地来。我们面临的那些问题是一点一点造成的，也许我们只得一点一点地慢慢解决。

污染是可以清理的。首先得从源头治起，就像清理垃圾那样，污染之源也可同时得到解决。通过鼓励或强制乱扔垃圾的人捡拾垃圾一段时间，便能较容易地改变其恶习。也许别的问题就没那么容易解决了，但这并不意味着它们是无法解决的。

我曾梦想过每栋大楼、每家工厂、每台车辆均有其自带的无污染能源。这该怎样实现呢？我不是太清楚，但是我知道大大小小依照具体情况定制的设备可以实现这一梦想。这种设备可以是某种转矩发电机，这种发电机能产生数以百计或者数以千计个使其运转所需的单位功率。我知道这类设备不是天方夜谭，且早已问世。千万别告诉我，说这是不可能的，因为我知道这已经存在。要说这不可能，全因人们总是不相信新事物的习惯、巨额利益、政府缺乏意愿使然。

Maybe such devices will come in other forms that harness gravity, hydrogen, nuclear fusion or something else that I am not even aware of yet. It is highly likely that when the will to allow and encourage these alternatives exists on a national and international level, there will be many options that we can use.

I personally think that population growth should be limited just as China has done and maybe even better. I respect the political will and the reasons behind this policy in China. Perhaps the world's population can continue to grow indefinitely, but the land we use to grow food to feed ourselves is finite. Maybe the resources that we have at our disposal really are infinite, but we are not yet clever enough to even feed ourselves without the anomalies of great food shortages and such an overabundance that in some countries the obesity problems are huge. (That's my attempt at a joke).

It's up to you to fix it. It's up to us all.

也许这类设施会以可控制重力、氢、核聚变的其他形式面世，只是我还没有意识到而已。一旦这些替代品被允许和鼓励在国家以及国际范围内使用，我们极可能会有很多方案可供采用。

我个人以为，人口增长应该像中国那样受到限制，也许还可能做得更好。我尊重中国计划生育政策的政治意愿及其背后的种种原因。也许地球的人口会继续无限制地增长下去，可我们用以种植粮食自给的土地是有限的。也许可供我们处理的资源是无限的，可惜我们还不够聪明，不能靠食物的异常匮乏和某些国家因过度充裕而致肥胖症来当饭吃（我试图开个玩笑而已）。

解决这个难题得靠你，靠我们所有人。

CHAPTER 13

Honours

I have received honours from the Wenzhou people that I could never have foreseen. For some reason my actions aroused a level of interest out of all proportion to what I have done. The most recent at this time in April 2014 is that a shelter building on the mountain has been named after me using my Chinese name 伊凡(yī fán)。

Many reporters have asked me how I feel about all these honours and the publicity that has happened. For a long time I just felt confused and puzzled by it all. I think that over time my feelings have steadied into something like the following:

I definitely feel honoured, and I am proud and happy to have been of service to the Wenzhou people and to China, BUT, I do not think that I deserve such honour because I know that I have NOT been totally altruistic in my motivations when I started picking up rubbish. Neither was I totally unselfish when I continued. Wenzhou in China are a part of my world and I like living in a beautiful world that is not polluted. I love beautiful mountains wherever they are on our little home (planet Earth), and other beautiful places as well. I think we should all do something to keep these places beautiful and I am willing to do

第十三章

荣 誉

我从温州人民那儿获得了自己不曾预知的种种荣誉。由于某种原因，我那些行动引起了关注的程度，远超出我所做的一切所应得的。最近我得到了一份殊荣，即 2014 年 4 月，大罗山上有个歇脚的亭子用我的中文名伊凡(yī fán)命名了。

不少记者都问我如何看待这些荣誉和已然确立的知名度。这一切让我茫然不知所措了很长时间。我想，过了一段时间，我的心情才渐渐平静下来，具体感受如下：

我自然感到非常荣幸，我为自己能服务温州人民乃至中国而感到骄傲和幸福，**不过**，我认为自己还不配获此殊荣，因为当我开始捡拾垃圾的时候，我知道自己**并非**一直毫无私心，我坚持工作也并非完全出于无私。温州乃至中国都属于我生活的那个世界的一部分，而我则喜欢生活在一个没有污染的美丽世界里。在我们这个小小家园中，任何地方的美丽大山以及其他赏心悦目之所，我都

something to help. That is all. I do not wish to make anybody feel ashamed, but I am willing to challenge behaviours. However, I am only willing to do this by being a servant and being willing to lose face myself. I am not willing to confront people except with a smile and by continuing to offer this service. My promise to you and to myself is that I will continue to help wherever I am.

My heart leaps with joy that a building will remember a person who simply picked up rubbish. I am just a little embarrassed that it is me. I would feel happier if it was somebody else, but I will stand up and accept the responsibility that comes with the honour because I choose to do so for the sake of our environment. I promise to strive to live up to the expectations that come with the honours, but I am not sure what your expectations might be.

It is important to remember and look at the measure of my ability to lead in all that has happened. A prime example is that right at this moment I have 400 students and I have told them all what I do and why I do it. I have invited them to join me. Most of them have seen some of the publicity that has followed me. In spite of that and the direct incentive I give them of 1% extra for their final mark for each time they come and practice their English with me during each rubbish-picking venture, I have only ever seen 50 of these students with me on the mountain. Of these there are less than 10 who have come more than once. There are other people that come from many different places but they too, rarely stick at the job for more than a few times. Am I still a leader? Sure, but I am definitely no natural leader

非常热爱。我认为我们大家都应该行动起来,让这些地方一直美丽下去,我非常愿意献出自己的一点绵薄之力。仅此而已。我不想让任何人难堪,但却愿意挑战一些行为。然而,我只愿意以仆人身份和不怕丢脸的方式去做这一切。除了一笑置之且继续尽力而为外,我不愿冒犯他人。我对你和我自己的承诺就是,不管我在哪里,我将继续努力。

为记住一个捡拾垃圾的人而造一座建筑,这让我的心洋溢着快乐。只是让我有些不好意思的是,那人竟然是我。倘若那是别人,我会更快乐,不过,我会站直身躯,接受随荣誉而来的那份责任,因为为了我们的环境,我愿意那样做。我承诺一定竭尽全力以不辜负随种种荣誉而来的种种期望,只是我还不太确定你可能会有哪些期望。

重要的是,看看并记住我以一己之力带头所产生的结果。一个典型的例证就是,就在这一刻,我有四百名学生,我已经将自己的所作所为以及原因一一告诉了他们。我已邀请过他们加入我的行列。他们当中的多数人对尾随我而来的知名度有所耳闻。尽管如此,他们每次来捡拾垃圾并同我练习英语口语,我就会给他们期终成绩1%的加分,可在山上我却只看到了五十名学生。就这群学生中,跟我去过山上一次以上的,不足十名。还有来自不同地方的其他人,可坚持几次的人很少见。我还是一名领导者吗?当然,不过,我显然不是一位天生的领导者,或者说不是一位拥有独特个

or one with a particularly attractive personality. Just in case you think I am being overly self-deprecating, let me assure you that my ego is in fine form. I am doing very well and thank you.

There have been other honours, trophies, mementos, souvenirs, and TV appearances. Out of respect and in remembrance of all the people who have been patiently working in different ways to help our environment without such recognition, I dedicate them all to you. I particularly dedicate them to those volunteers who have come back with me week after week. You have encouraged me and inspired me to keep going. Now there is more work to do so let's keep at it. There may be no more formal recognition of our efforts but the work still needs to be done. We may receive even more honours and awards but the work still needs to be done. We may get tired and disillusioned, we may get sick and someone close to us will die but the work still needs to be done. Doing the work, can be our best reward.

性魅力的领导者。假如你认为我妄自菲薄，那我向你保证，我对自己有清醒的认识。我很好，谢谢。

我还获得了一些别的荣誉、奖品、纪念品以及电视露面机会。出于尊重，且为纪念所有那些一直以各种方式孜孜不倦且默默无闻为改善环境而行动的人，我将自己所有的荣誉献给诸位。我将它们特别献给那些每周都与我在山上工作的志愿者们。是你们鼓励且鞭策我继续坚持下去。现在还有更多的工作要做，就让我们不断努力吧。我们的努力可能不会再得到正式的认可，但工作还需要做下去。我们也可能会得到更多的荣誉和奖励，工作也一样需要去做。我们可能会疲倦，可能会灰心失望，我们可能会生病，身边亲近的人会死去，但工作还需要去做。做事可以成为我们最好的回报。

CHAPTER 14

To an Unknown Administrator

I wrote the following just prior to leaving Wenzhou Vocational and Technical College：

Wenzhou Vocational and Technical College—Internal Memorandum

From：Ivan Davis，Foreign English teacher. Mechanical Engineering Department

To：An Unknown Administrator

Subject：Recognition and gratitude

I have been recognised as doing something worthwhile for our environment by going to Da Luo Mountain each week to pick up rubbish，but you have never been recognised and possibly never will.

It is because of your decision to have picking up rubbish with me on Sunday mornings，included as part of the organised volunteer activities for students，that this activity has been such a success. Without the participation of these groups of students joining me almost every Sunday morning，my meagre efforts would have had little

第十四章

致无名的管理者

以下这些,是我离开温州职业技术学院前写下的:

温州职业技术学院内部备忘录

来自:机械工程系外籍英语教师　戴爱文

致:无名的管理者

主题:认可及感谢

我每个星期去大罗山捡拾垃圾的努力已得到认可,被认为是对我们的环境有益的行为,但你却没有得到认可,也许永远也不会得到认可。

你每周日与我同去大罗山捡拾垃圾的这一决定,是有组织的学生志愿者活动的一部分,正因为如此,我们的这一行动才取得了成功。如果没有这些学生每个周日早上与我同行,我个人那点微

impact on the environment and the community.

Your action has directly and positively affected thousands of students and possibly millions of people. In spite of this, you may never be recognised. I hope this memo will remedy this situation.

In six weeks, sadly I will be leaving China, so I want to make this opportunity to thank you and recognise you. If it becomes possible to do this in person, I would be delighted.

Ivan Davis
Phone: 150 6785 8141
Email: davis1953@westnet.com.au
May 14, 2013

不足道的努力对环境和社区将不可能产生什么影响。

你的行动直接而积极地影响了成百上千的学生，甚至可能有成千上万的市民。尽管如此，你的行动也许从来不会得到认可。我希望这份备忘录能够弥补这一憾事。

遗憾的是，六个星期后我将离开中国，因此想借此机会对你表示认同和谢意。如果能当面致谢，我将荣幸之至。

戴爱文（伊凡）
电话：150 6785 8141
电子邮箱：davis1953@westnet.com.au
2013 年 5 月 14 日

CHAPTER 15

Final Thanks

There have been many times in my life where success has eluded me. It was during those times that I never sought advice because mostly my attitude was that I knew everything I needed to know. It was during those times I found it difficult to be thankful. Now, it is with pleasure and a great deal of humility that I offer the following thanks.

First of all I want to say thanks to that power in the universe that gave me a hand up when I asked for it. I had to get to a pretty bad place before I asked for help, but there you go. It seemed like a miracle to me at the time but really the only miracle was that I finally put away my stubborn pride and asked for help. The help was always there just waiting for me to ask for it.

Thanks to my family who loved me unconditionally and have always supported me even when I was doing pretty crazy things. A special thanks to my mother who has been a constant inspiration. She wrote her first book at the age of 80 and in doing so took away all excuses from me and from many other people for our lack of action. Now at the age of 90 she is still busy creating, supporting others, and working at different projects even though she invariably can only work for half of any given day.

第十五章

最后的致谢

在我一生中，许多事情都以失败告终。在每一次失败中，我从未求助过别人的指点，因为我大都持有自以为该知道的都知道的态度；在每一次失败中，我发现向他人表示谢意都是一件难事。现在，我以愉悦以及极大的谦卑之情，表达如下诸多感谢。

首先，我想感谢上苍在我需要力量时给予了及时的援手。在我只得茫然前行时，我想到了寻求帮助，瞧！你来啦。在我当时看来，这简直是个奇迹，但事实上，这唯一的奇迹是我最终收起了自己那顽固的自尊，开始求助于人。其实，援手一直在那儿，就等着我去寻求而已。

感谢我的家人给予我无条件的爱，即便是在我做出一些疯狂之举时仍支持着我。特别感谢给予我源源不断灵感的母亲。她在八十高龄时写出自己的第一本书，她的这一举动，使得我和许多人再也无法为自己的无为找到任何借口。现在，年过九旬的她仍然忙于创作、援助他人，且同时着手不同的项目，尽管她每天只能工作半天的时间。

Thanks to Angela who, even though she is a traditional girl in so many ways, never expected me to be home doing the traditional husband thing. Without the freedom she has allowed, many of my adventures would never have happened.

Thanks to Bill Harris and www. centerpointe. com for the great meditation and life skills programme. It is only because of the increased awareness, knowledge of my personal power, and a lot more effective neural connections operating in my brains, which have enabled me to have a myriad of adventures including going to China and writing this booklet.

Thanks to David Xiong and the other staff from the Foreign Affairs Department of Wenzhou University for recommending me to Oujiang College. You supported me admirably in my attempts to return to Wenzhou. A special thanks must go to Mr. Ma as head of Oujiang College because he has not only helped with the publication of this book, he also has my boss Ms. Peng Xiaoying as one of his staff. She is not only a wonderful boss but she was able and willing to take action, to interview me and then hire me when it seemed that nobody wanted me back in Wenzhou. While other people are procrastinating, she follows through quickly with what she says she will do. Thanks boss!

Finally thank you to China for welcoming me so readily even with my crazy ways. Thank you to those wonderful Chinese people who have befriended me and loved me as I am, even with my plethora of faults. You welcomed me into your homes and into your hearts. I have been greatly honoured. In particular thank you to the people of Wenzhou and Chashan. My eyes well up

感谢安吉拉，尽管她在诸多方面是个传统的女性，但却从未指望我做一个传统的丈夫。若没有她给予我的那份自由，我不少的冒险活动就不可能实现。

感谢比尔·哈里斯及 www.centerpointe.com，两者为我提供了无尽的冥思机会及生活技巧项目。我越来越意识到并且明白自己的个人能力，以及在我脑子中起作用的中枢神经之间的联系越来越多，这两者使得我有勇气不揣冒昧，经历无数的冒险，这其中就包括去中国及写下这本小册子。

真诚地感谢熊勇及温州大学外事处将我推荐给瓯江学院。正是你们那可贵的支持，我才圆了重返温州之梦。在此我要特别感谢瓯江学院马光院长，不仅因为他帮忙出版本册子，而且还让其员工彭小英女士领导我。彭女士不仅是一位了不起的领导者，而且非常能干，说干就干，在别人未必愿让我重返温州时，是她面试并雇用了我。在别人拖延之时，她迅速完成了她承诺的事。多谢领导！

最后，感谢中国。感谢她乐意接纳我这个疯狂的人。感谢那些了不起的中国人，我有着数不尽的缺点，感谢他们对我这个满身都是缺点的人的友好和善待。是你们让我走进你们的家中以及你

with tears when I start to think of all you have given me. So many individuals have given so much to me even when there was nothing I was giving to them. That type of giving is true giving and it has inspired me to keep giving myself. Thank you, thank you, thank you!

们的心中。我感到荣幸之至。特别感谢温州人民，尤其是茶山人民。每当我想起你们大家所给予我的一切，我便会热泪盈眶。即便我无以回赠，确有那么多人给予了我太多太多。这种给予才是真诚的给予，正是这种给予一直激励着我不停地奉献自己。谢谢你们！谢谢你们！谢谢你们！

本书的文字，在交予出版社之前，由曹顺发校对，在此表示感谢。

Special thanks go to Professor Cao Shunfa who has done proofreading for this book.